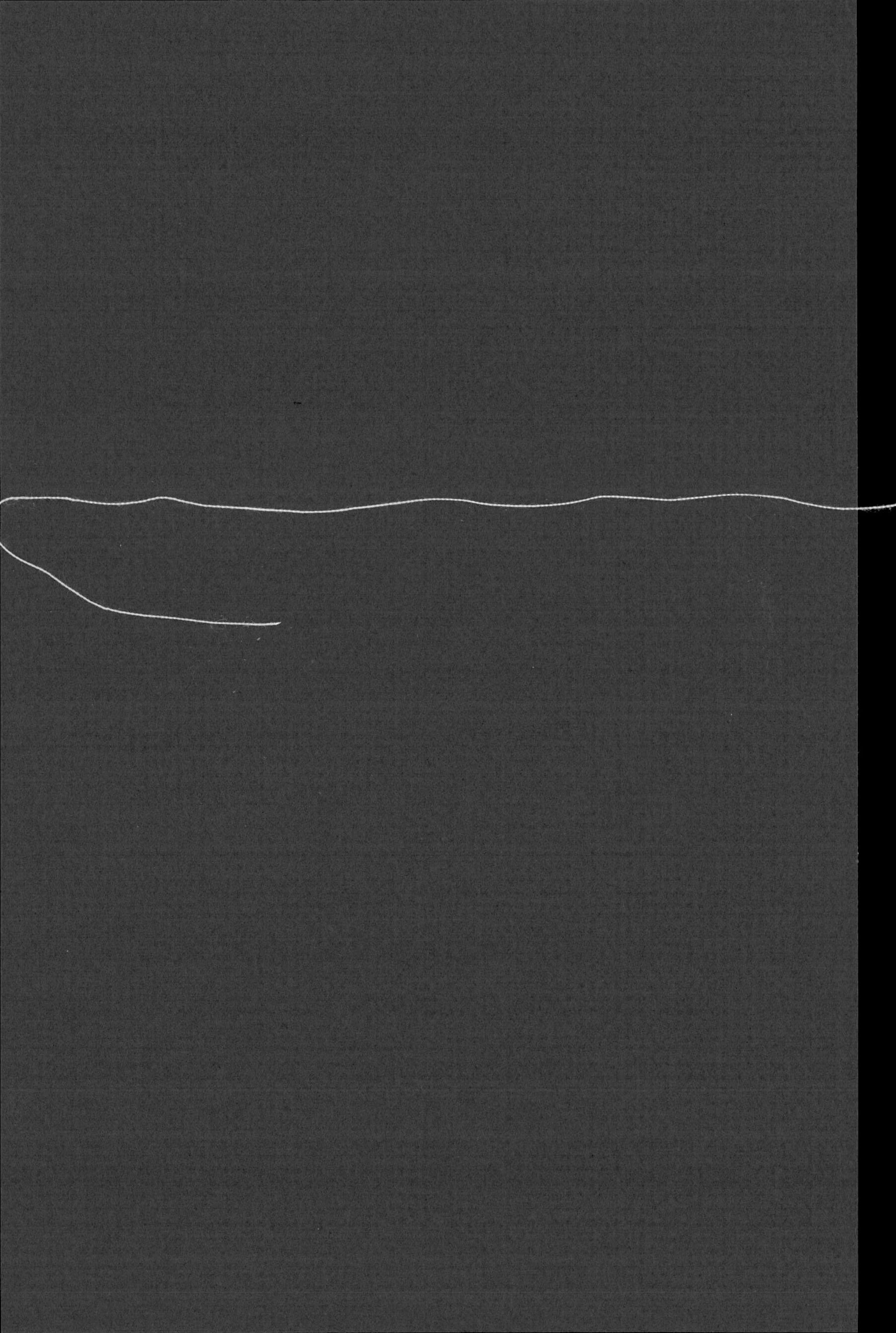

field guide to the Vascular Plants of Grand Teton National Park and Teton County, Wyoming

field guide to the Vascular Plants of Grand Teton National Park and Teton County, Wyoming

Richard J. Shaw

with major contributions by Robert D. Dorn

Utah State University Press
Logan, Utah

Library of Congress Cataloging in Publication Data

Shaw, Richard J
Field guide to the vascular plants of Grand Teton National Park and Teton County, Wyoming.

Bibliography: p.
1. Botany — Wyoming — Grand Teton National Park. 2. Botany — Wyoming — Teton Co. 3. Grand Teton National Park. I. Title.
QK195.S5 581.9'787'55 75-26537
ISBN 0-87421-081-X

For Marion

Special Acknowledgement

The Grand Teton Natural History Association, a non-profit organization, gave money to aid in the publication of this book. This association, established in 1933, is dedicated to support historical, scientific and educational activity of the National Park Service.

Contents

Preface

The preparation of this handbook culminates the work of 25 years of association with Grand Teton National Park as a seasonal naturalist. This association has allowed the author to work with numerous taxonomists, ecologists, high school and university students, and hosts of eager park visitors. From conducted walks, campfire programs and serious field studies over this span of years, the author is convinced that a comprehensive and yet lucidly written handbook can treat the needs of a broad spectrum of users interested in the flora of this majestic mountain landscape. The author also strongly believes after 18 years of teaching taxonomy to undergraduates at Utah State University, that the indented dichotomous key can make the identification of an unknown plant as easy as possible. Every effort has been made to construct keys which use a minimum of technical, taxonomic terminology.

The vascular plant species treated in this book are growing wild and have been compiled largely from published sources based on thousands of herbarium specimens scattered throughout the United States. The Rocky Mountain Herbarium at the University of Wyoming, the Intermountain Herbarium at Utah State University and the Herbarium at Grand Teton National Park have been consulted most frequently. Two annotated checklists published 10 years apart (Shaw 1958-1968) have served as the immediate base for this flora, but many interested people have contributed valuable data on newly discovered taxa. It is impossible to create such a handbook and not have qualms about insufficient time to study difficult or critical genera. The author is acutely aware that different taxonomists do not entirely agree as to classification, and so, as in many manuals, this book is uneven in its quality from group to group. No claim is made for the completeness of the treatment of species. Therefore, the author will welcome suggestions and information about additional vascular plants of the area.

Every other botanist may aspire for praise, but the plant taxonomist can only hope to escape reproach.

Samuel Johnson
1709-1784

Explanations and Arrangement of the Flora

Families are arranged alphabetically within divisions or classes. A similar arrangement is followed for genera within families and species within genera. Alien species are preceded by the symbol "†". In most cases an attempt has been made to include the abundance of each species and its usual habitat. Rare, infrequent, frequent, common and abundant are the relative terms used.

Synonomy is not complete, and includes only those synonyms which are most apt to be used in recent taxonomic references. Common names, when available, have been used in connection with the generic description and in a few cases unique common names have been used in the species listing. In the majority of cases the nomenclature and taxonomy follow Hitchcock and Cronquist with supplementation from other taxonomists listed in Selected References (page 288).

Using the Key

The key used in this flora is of the indented type, and is dichotomous (twice branching) throughout. This means that at every stage a choice must be made between two contrasting alternatives. As the main key allows for the identification of 88 families, it has been arranged in some cases into groups with a group key first. To speed up reference to particular leads (alternative statements) each pair of leads is numbered.

To find the family to which a specimen belongs, one starts by comparing the specimens with the first two alternatives numbered "1" in the key. If the specimen agrees with the first alternative, one proceeds to the pair of alternatives numbered "2". If it agrees with the second alternative, one then proceeds to the pair numbered "3". The process is continued until a family name is reached. It is very important that at each step the entire lead is considered carefully; otherwise the wrong identifications may be made.

It will sometimes happen that the specimen does not agree with all the characters given in a particular alternative. When this condition is recognized, one must decide which of the two leads fits the specimen best. In general, the most reliable characters are placed at the beginnings of the leads, so these characters should be observed most carefully.

Difficulties in Using Keys

Botanical keys are not perfect, but are like the people that create them. Below are listed some difficulties encountered by students.

1. The key may be based upon the average plant. A specimen may be abnormal in some way. Sample the plant population and select one or more average plants.
2. The key may depend upon characters not present on the plant. For example the generic key to the family Brassicaceae (Mustard Family) is usually based on the fruit type, but features of the flower are necessary to key the plant to the family. If the plant does not have both flowers and fruit, one may have to wait a few days for fruit maturation.
3. The authors may have created an ambiguous or faulty portion in the key. This unfortunate situation can often be corrected by an instructor. If one is on his own, all he can do is to check the plant again, or try other parts of the key. If one continues to come back to the same place, one might suspect an omission or an error in the key.
4. The plant to be identified may have been left out of the key. Weedy species are continually migrating and may even have become locally common since the flora was written. It may also happen that the plant may be rare or perhaps the limits of the species were not understood at the time the flora was completed. When this may be the case, try a flora or manual from an adjacent area.

Aliens in the Flora

The word "alien" in this handbook is used to refer to those plants that have migrated directly into the park or county within the last 75-100 years, or those plants remaining from cultivation usually around abandoned places of habitation. The status of such aliens is difficult to determine. Some taxonomists feel that a flora should include established aliens, but exclude those recorded only once or twice. The author feels, however, that in view of the tremendous visitation to G.T.N.P. (many years exceeding 3 million visitors) that an attempt should be made to document as many aliens as possible. Changes in the status of aliens have occurred, and their role in the flora should be periodically reevaluated. Comparison of the 1968 checklist with this handbook will point to an increasing "pollution" of the park flora with aliens, some of which have already become serious weeds. In 1968, 72 aliens were noted in Shaw's checklist. In this volume 88 aliens are recorded (the symbol "†" precedes a known alien in species listing).

Acknowledgements

It is difficult to give credit to all who have helped in one way or another in the preparation of this handbook. My gratitude for assistance given is extended to those persons specifically named and to the many I do not name. Among those who should be especially recognized are: Robert D. Dorn, not only for his treatments of Poaceae and the genera *Artemisia* and *Salix*, but also for his tireless effort in pointing out taxa which had been missed in previous searches; Arthur H. Holmgren for much assistance over many years in the identification of collected specimens; Jack Major for his stimulating ecological observations and urgings to complete the project. Barbara Houghton has typed most of the manuscript and my wife, Marion Shaw, has done a great deal of the editing and correcting of the manuscript.

I feel special gratitude toward those groups who have given me money for the publication of this book. Especially, of course, I am grateful to Grand Teton Natural History Association, but additionally I am grateful to the Biology Department of Utah State University for the financial aid given to me as well as the encouragement given by Gene W. Miller, head of that department.

Statistical Summary

STATISTICAL SUMMARY OF THE VASCULAR PLANTS OF G.T.N.P. AND TETON COUNTY, WYOMING

	Indigenous			Alien Species		
	Family	Genera	Species	Family	Genera	Species
Pteridophytes	7	19	31	0	0	0
Gymnosperms	2	5	9	0	0	0
Dicotyledons	58	232	563	7	38	75
Monocotyledons	14	71	233	0	2	10

Grand Totals:

Families — 88
Genera — 367
Species — 921

Largest Families (Native + Alien Species)

Asteraceae	111 + 22
Poaceae	114 + 10
Cyperaceae	48 + 0
Brassicaceae	34 + 10
Scrophulariaceae	41 + 3
Rosaceae	35 + 2

Largest Genera (Native + Alien Species)

Carex	37 + 0
Poa	22 + 2
Salix	20 + 1
Artemisia	13 + 1
Ranunculus	13 + 0
Potentilla	12 + 0
Aster	12 + 0
Erigeron	11 + 0
Juncus	10 + 0

Abbreviations

alt	alternate
bisex	bisexual
Ca	calyx
Co	corolla
fl	flower
infl	inflorescence
lvs	leaves
opp	opposite
ov	ovary
per	perianth
P	pistil
reg	regular
S	stamens
stip	stipules
unisex	unisexual
*	parts are united
∞	parts are more than 10
/	or
( )	sometimes
X	few
†	alien species

field guide to the Vascular Plants of Grand Teton National Park and Teton County, Wyoming

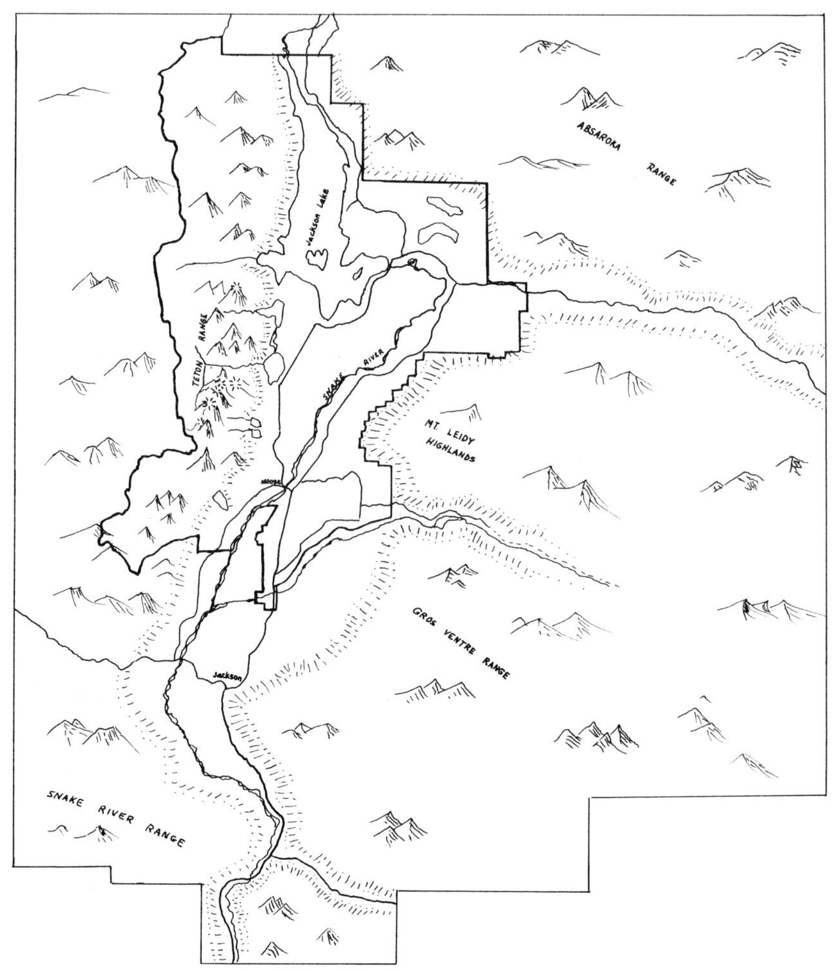

Teton County, Wyoming

Plate 1

Snake River, Jackson Hole and Teton Range

Fireweed, Butter-and-Eggs

Subalpine fir krumholz on one side of boulder

Plate 2

Subalpine fir replacing lodgepole

Balsamorhiza sagittata Arrowleaf Balsamroot

Erysimum asperum Wallflower

Plate 3

Nuphar polysepalum Spatterdock

Pseudotsuga menziesii Douglas Fir

Wyethia helianthoides Mulesears

Plate 4

Iliamna rivularis Globemallow (by Margaret Shaw)

Silene acaulis Moss Campion (by Margaret Shaw)

Eriophyllum lanatum Eriophyllum

Plate 5

Lonicera utahensis Utah Honeysuckle

Castilleja rhexifolia Rhexia-leaved Paintbrush

Mimulus lewisii Lewis Monkey Flower

Plate 6

Sorbus scopulina Mountain Ash

Kalmia microphylla Swamp Laurel

Amelanchier alnifolia Serviceberry

Plate 7

Primula parryi Parry's Primrose

Pteridium aquilinum Bracken Fern

Marsilea vestita Pepperwort

Plate 8

Ceanothus velutinus Ceanothus

Erythronium grandiflorum Glacier Lily

Pedicularis groenlandica Elephant Head

Plate 9

Actaea rubra Baneberry

Fragaria vesca Strawberry

Caltha leptosepala Marshmarigold

Plate 10

Aconitum columbianum Monkshood

Potentilla gracilis Cinquefoil

Saxifraga flagellaris Whiplash Saxifrage

Plate 11

Gilia aggregata Skyrocket Gilia

Ranunculus eschscholtzii Alpine Buttercup

Saxifraga oppositifolia Alpine Saxifrage

Key to the Divisions and Classes

1 Plants not producing seeds; spores are the usual structures for disseminating the plants
  2 Plants with microphylls (leaves small with single midvein); sporophylls or sporangiophores often aggregated into a terminal cone
    3 Leaves free, mostly alternate or opposite; stems not jointed .... Division I Lycopodiophyta, P. 2
    3 Leaves connate and whorled; stems longitudinally ribbed, hollow and jointed ....Division II Equistetophyta, P. 2
  2 Plants with macrophylls (large leaves in most species, usually with more than a single vein); leaf gaps present; plants lack cone aggregations of sporophylls .... Division III Polypodiophyta, P. 2
1 Plants producing seeds; seeds are the usual structures for disseminating the plants
  4 Ovules naked, and borne on the surface of ovuliferous scales which are crowded together into a cone; flowers absent .... Division IV Pinophyta, P. 2
  4 Ovules enclosed in an ovary; flowers present .... Division V Magnoliophyta,
    5 Flower parts in 4's or 5's; leaves usually with reticulate venation; cotyledons 2 .... Class Magnoliopsida (Dicots), P. 23
    5 Flower parts usually in 3's; leaves usually with parallel venation; cotyledons 1 .... Class Liliopsida (Monocots), P. 31

Key to the Families

Division Lycopodiophyta

Key to the Families

1 Plants aquatic, often submerged in ponds or occasionally growing on exposed mud; sporangia at the base of long slender leaves .... Isoetaceae, P. 5
1 Plants terrestrial; leaves small and scale-like
  2 Plants homosporous; terminal strobilus round or lacking .... Lycopodiaceae, P. 5
  2 Plants heterosporous; terminal strobilus usually 4-sided .... Selaginellaceae, P. 6

Division Equisetophyta

The division has only one living family .... Equisetaceae, P. 6

Division Polypodiophyta

Key to the Families

1 Plants terrestrial (though often in moist habitats); plants homosporous; sporangia borne on leaves, not enclosed in sporocarps
  2 Sporangium opening by a transverse slit, no annulus; leaves sometimes bent over in bud, but not circinate .... Ophioglossaceae, P. 9
  2 Sporangium opening by means of an annulus; leaves circinate in bud .... Polypodiaceae, P. 10
1 Plants aquatic or semiaquatic; plants heterosporous; sporangia borne in sporocarps .... Marsileaceae, P. 9

Division Pinophyta (Gymnospermae)

Key to the Families

1 Leaves scale-like or somewhat needle-like, in whorls or opposite; seed cones small with few scales (2-12) .... Cupressaceae, P. 15
1 Leaves needle-like, single or in clusters of 2-5; seed cones woody with many scales .... Pinaceae, P. 15

Key to Genera and Species of Ferns, Fern Allies and Gymnosperms

ISOETACEAE Quillwort Family

Aquatic, perennial herbs with tufted annual, quill-like leaves arising from a submerged fleshy axis; spores of two types borne on the adaxial surface of the leaves.

ISOETES Quillwort

A single genus with the characteristics of the family.

1. **I. bolanderi** Engelm. Rare in shallow water of ponds near Colter Bay.

LYCOPODIACEAE Clubmoss Family

Perennial herbs with small eligulate leaves, less than 1 cm long; spores all alike and numerous; sporangia axillary; sporophylls resembling vegetative leaves.

LYCOPODIUM Clubmoss

A single genus with the characteristics of the family.

1 Sporophylls resembling vegetative leaves .... **L. selago**
1 Sporophylls obviously different from vegetative leaves .. .. **L. annotinum**

1. **L. annotinum** L. Rare, in deep shade on the west sides of Jenny and Leigh Lakes, circumboreal.
2. **L. selago** L. Rare, occuring in cliffs. Only one collection.

SELAGINELLACEAE Selaginella Family

Plants low and creeping, moss-like in appearance; leaves and sporophylls small and numerous not more than 5 mm long, ligule at the base; spores of two kinds in a 4-angled terminal strobilus.

SELAGINELLA Spikemoss; Selaginella

A single genus with the characteristics of the family.

1 Strobili not 4-angled; leaves soft and thin, without a dorsal groove; rare .... **S. selaginoides**
1 Strobili 4-angled (square in x-section); leaves thick and firm, dorsal groove present; widespread .... **S. densa**

1. **S. densa** Rybd. Locally common on exposed, rocky sites in the valley and sometimes extending above timberline.
2. **S. selaginoides** (L.) Link. Rare, Treasure Lake, Teton Canyon.

EQUISETACEAE Horsetail Family

Perennial herbs with rhizomes; stems cylindrical, jointed and mostly hollow; green foliage lacking, but bearing minute scarious leaves in a whorl at the nodes; terminal strobilus composed of peltate sporangiophores each bearing 5-10 elongate sporangia under a polygonal cap.

EQUISETUM Horsetails and Scouring Rushes

Only one genus with the characteristics of the family.

1 Aerial stems annual, often of 2 kinds, usually some of them with whorls of branches; strobili blunt, not apiculate; minute leaves of sheath persistent (horsetails).
  2 Fertile and sterile stems alike; sporangia produced in summer .... **E. fluviatile**
  2 Fertile and sterile stems obviously different; sporangia produced in spring .... **E. arvense**
1 Aerial stems perennial and evergreen (except in *E. laevigatum*); minute leaves of sheath often deciduous (scouring rushes)
  3 Stems small and slender, mostly 1-3 dm tall, 5-12 ridges on stem .... **E. variegatum**

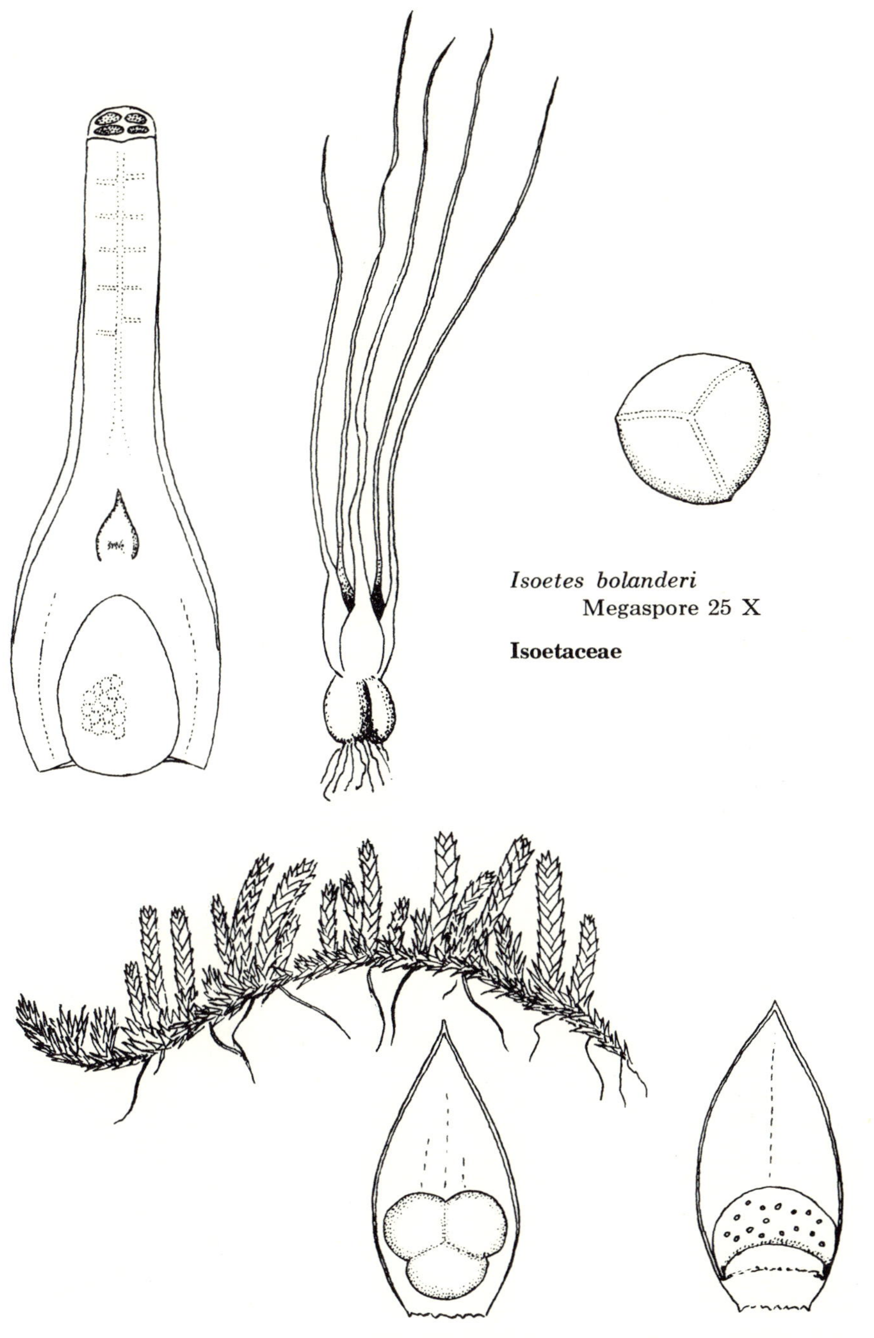

Isoetes bolanderi
Megaspore 25 X

Isoetaceae

Megasporophyll 10X   Microsporophyll 10X

Selaginella densa Selaginella   **Selaginellaceae**

3 Stems tall and robust, mostly 2-15 dm tall, 16-40 ridges on stem
- 4 Aerial stems annual; strobili blunt or inconspicuously apiculate .... **E. laevigatum**
- 4 Aerial stems perennial; evergreen; strobili distinctly apiculate .... **E. hyemale**

1. **E. arvense** L. Horsetail. Cosmopolitan; locally common in moist areas of the valley.
2. **E. fluviatile** L. Horsetail. Marshes and bogs. Only one collection made about 1900.
3. **E. hyemale** L. Infrequent in the area of the Oxbow Bend of the Snake River, circumboreal.
4. **E. laevigatum** A. Br. Common in the sedge meadows near the Oxbow Bend of the Snake River [*E. kansanum* Schaffn.].
5. **E. variegatum** Schleich. Infrequent in wet meadows and streambanks of the valley, circumboreal.

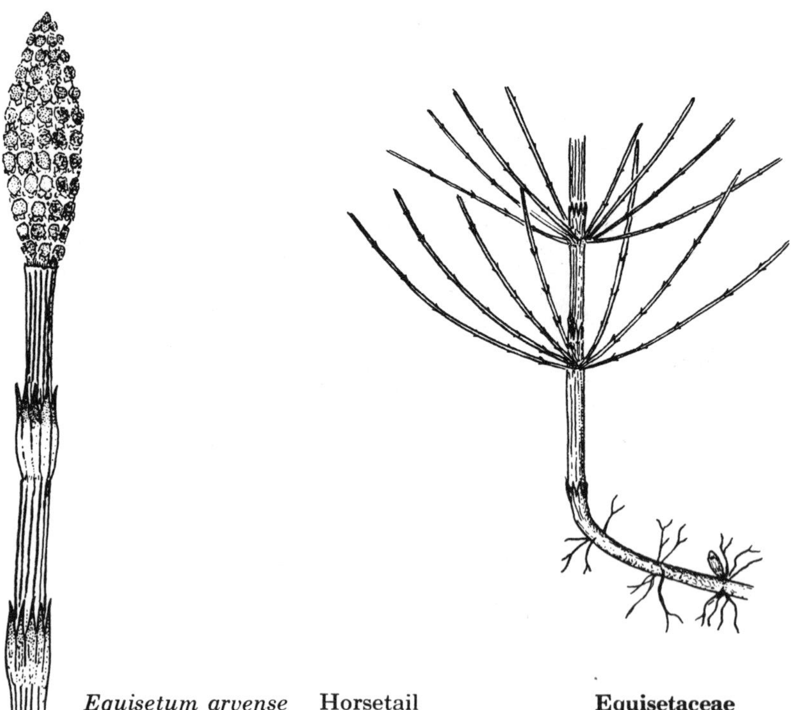

Equisetum arvense Horsetail **Equisetaceae**

MARSILEACEAE Pepperwort Family

Aquatic perennial with creeping rhizomes; leaves with long petioles, palmately 4-foliate; sporocarps borne on short stipes, hard and bony; heterosporous.

MARSILEA Pepperwort, Water Clover

A single genus with the characteristics of the family.

1. M. vestita Hook & Grev. Locally frequent in small ponds of the valley.

OPHIOGLOSSACEAE Adder's Tongue Family

Perennial, fleshy herbs with simple or compound leaves; erect underground stem; sporangia ovoid to globose without a true annulus, but opening by a transverse slit, each sporangium gaping at maturity.

BOTRYCHIUM Grape Fern

Plants growing on soil, fleshy roots, stem short, erect and fleshy. Single leaf per year with a common stalk bearing one

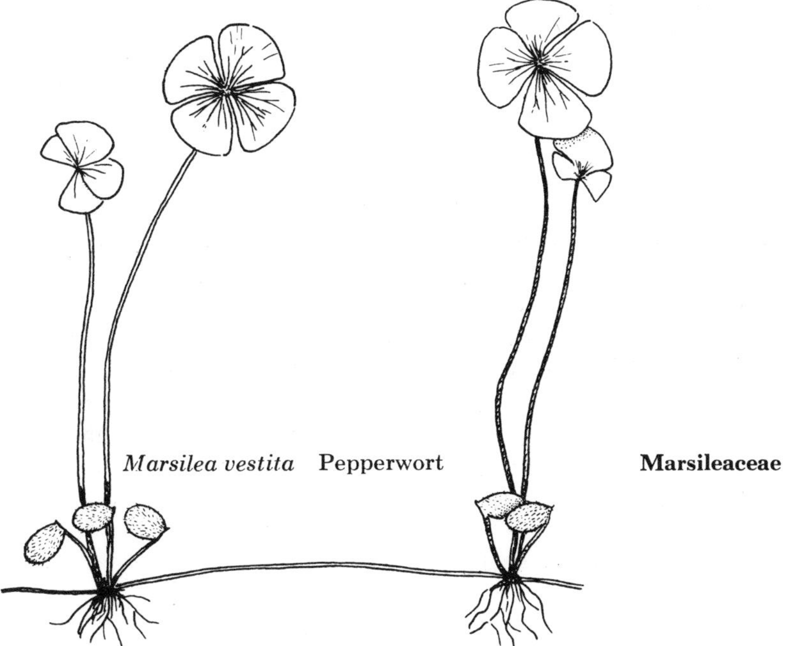

Marsilea vestita Pepperwort **Marsileaceae**

sterile blade and one fertile blade; sterile segment pinnately compound or dissected.

1 Sterile leaf blade deciduous, appearing nearly sessile and attached above the middle of the plant .... **B. virginianum**
1 Sterile leaf blade evergreen, on a well-developed petiole 1-12 cm long and attached near the ground level . . **B. multifidum**

1. **B. multifidum** (Gmel.) Trevis. Rare near the outlet of Jenny Lake.
2. **B. virginianum** (L.) Swartz. Vicinity of Boy Scout Camp in Teton Canyon, west of western park boundary.

POLYPODIACEAE Fern Family

Perennials with rhizomes; leaves with circinate vernation in bud; sori naked, covered by an indusium or by the recurved leaf margin; sporangium with an annulus; spores all alike.

1 Sori marginal or submarginal, protected by recurved leaf margin
 2 Leaves dimorphic, the fertile one longer than the sterile one
  3 Petioles greenish or greenish straw color . . **Cryptogramma**
  3 Petioles dark reddish brown .... **Aspidotis**
 2 Leaves monomorphic
  4 Sori attached to and covered by the thin, strongly reflexed tips of the leaflet lobes .... **Adiantum**
  4 Sori more or less continuous and covered by the recurved leaf margin
   5 Leaves woolly-hairy or scaly beneath, ultimate segments less than 5 mm long .... **Cheilanthes**
   5 Leaves glabrous or weakly hairy beneath, ultimate segments usually over 5 mm long
    6 Plants densely tufted, mostly growing in crevices or among rocks .... **Pellaea**
    6 Plants not densely tufted, leaves single, growing in soil .... **Pteridium**
1 Sori on the veins, not marginal, but sometimes submarginal, not protected by recurved leaf margin
 7 Indusia lacking (use hand lens) .... **Gymnocarpium**
 7 Indusia present
  8 Sori round in outline

9 Indusia arising from under the sori, (inferior)
10 Indusia attached at one side only, forming a hood and peeling backward at maturity .... **Cystopteris**
10 Indusia attached all around, splitting into pie-shaped segments from the top and center .... **Woodsia**
9 Indusia attached above the sori on a little hump, (superior)
11 Indusia attached at one side by the notch .... **Dryopteris**
11 Indusia shield-shaped, attached at the center and spreading over the sorus .... **Polystichum**
8 Sori elongated in outline
12 Leaves once compound .... **Asplenium**
12 Leave at least twice compound .... **Athyrium**

ADIANTUM Maidenhair Fern

Mesophytic ferns with slender creeping rhizomes. Lvs compound with widely spreading pinnae, petioles slender, dark and shiny; sori marginal covered by a false indusium formed by recurved tips of the leaflet lobes.

1. **A. pedatum** L. Infrequent along stream on south facing slope 2 miles up Waterfalls Canyon, also Death Canyon.

ASPIDOTIS Pod-fern

Small ferns of mesophytic, rocky habitats, short rhizomes. Lvs glabrous, evergreen, 2-4 times pinnate, petioles slender mostly brown, monomorphic or only slightly dimorphic; margins of the fertile pinnules reflexed.

1. **A. densa** (Brackenr.) Lellinger. Frequent in rocks on the south facing slope of Waterfalls Canyon. [*Cheilanthes siliquosa* Maxon.].

ASPLENIUM Spleenwort

Small evergreen fern with short, scaly rhizome. Lvs (in ours) once pinnate, elongate and narrow; veins not reaching the margins;

petiole wiry; sori elongate along the veinlet; indusium thin and flap-like, opening toward the midline of the pinna.
1. **A. viride** Huds. Rare, especially in limestone at high elevations.

ATHYRIUM Lady Fern

Ferns of medium size from scaly creeping rhizomes. Lvs tightly bunched up to 10 dm long or longer; blade 2-4 times pinnate; sori mostly elongate and curved across the veins; indusium thin and fragile, often difficult to observe, lacking in some species.

1 Indusium usually visible with a hand lens; veinlets evident on the lower surface of the pinnule .... **A. filix-femina**
1 Indusium lacking; veinlets obscure .... **A. distentifolium**

1. **A. distentifolium** Tausch ex Opiz var. **americanum** (Butters) Cronq. Frequent by small streams in the North Fork of Cascade Canyon. [*A. alpestre* (Hoppe) Rylands].
2. **A. filix-femina** (L.) Roth. Frequent along streams draining into the valley lakes, also up to 7,500 ft. in the major canyons.

CHEILANTHES Lip-fern

More or less xeromorphic ferns less than 25 cm high; rhizome short, beset with brown to blackish scales. Lvs evergreen, pinnate, more or less woolly or scaly or both; sori borne on vein ends, submarginal, the margins somewhat reflexed or inrolled.
1. **C. feei** Moore. Rare, rocky sites at low to middle elevations.

CRYPTOGRAMMA Rock-brake Fern

Ferns mostly small, densely tufted from short thick rhizomes. Lvs of two kinds; sterile ones with broad, flat segments; fertile ones much longer than the sterile; mostly 2-3 times pinnate; margins of fertile pinnules reflexed to form a continuous false indusium.

1 Fronds mostly scattered on an elongate slender rhizome; petioles brown to dark purple at the base .... **C. stelleri**
1 Fronds densely crowded on a short branched rhizome; petioles greenish or straw-colored .... **C. crispa**

1. **C. crispa** (L.) R. Br. ex Hook var. **acrostichoides** (R. Br.) Clarke. Frequent in rocks, lower portions of the major canyons.
2. **C. stelleri** (Gmel.) Prantl. Rare, usually wet limestone.

CYSTOPTERIS Bladder Fern

Small to medium-sized ferns; short, creeping rhizome. Lvs variable in size, 2-4 times pinnate; sori borne on veinlets which continue to the margin, indusium thin and translucent forming a hood which pulls back as the sorus expands, disappearing early.

1. **C. fragilis** (L.) Bernh. Frequent in rock crevices of the canyons.

DRYOPTERIS Shield Fern; Wood Fern

Medium-size ferns from scaly rhizomes, covered with chaffy scales, and often persistent petiole-bases remaining from previous years. Lvs deciduous or evergreen, blade 2-3 times pinnate; sori roundish, each borne on a vein, indusium kidney-shaped or horse-shoe shaped, attached by a deep notch on one side.

1. **D. assimilis** Walker. Moist shaded woods, mouth of Cascade Creek.

GYMNOCARPIUM Oak Fern

Mesophytic ferns with slender somewhat scaly rhizomes. Lvs glabrous or glandular, at least 3 times compound; sori round or broadly elliptic, positioned on the veins, submarginal, indusium lacking.

1. **G. dryopteris** (L.) Newm. Rare, moist wooded areas.

PELLAEA Cliff-brake Fern

Small tufted ferns growing in dry habitats, from short branched rhizome covered with brown scales. Lvs evergreen, 1-4 pinnate, petioles green to reddish-brown; sori borne on the ends of veins just within the revolute margin which forms a false indusium.

1 Persistent petiole-bases usually more numerous than the green fronds; petioles with a series of basal, cross grooves; lower pinnae usually bifid .... **P. breweri**

1 Persistent petiole-bases usually fewer than the green fronds; petioles mostly without grooves; pinnae seldom bifid .... **P. glabella**

1. **P. breweri** D. C. Eat. Among rocks, usually limestone, at low to high elevations.

2. **P. glabella** Mett. var. **occidentalis** (E. Nels.) Butters. Infrequent, cliff crevices, north end of Blacktail Butte. [Treated by some as *P. occidentalis* (E. Nels.) Rydb.].

POLYSTICHUM Holly Fern

Mesophytic, tufted ferns from scaly, stout rhizomes. Lvs coarse and evergreen, pinnate, pinnae with small teeth; sori on the veins, usually in one or more definite rows; indusium arising from the middle of the sorus, centrally attached, margin usually fringed.

1 Lower pinnae of the frond divided and deeply cleft; plants up to 3.5 dm high .... **P. scopulinum**
1 Lower pinnae of the frond undivided; plants up to 6 dm high .... **P. lonchitis**

1. **P. lonchitis** (L.) Roth. Frequent in talus slopes and crevices in major canyons and couloirs.
2. **P. scopulinum** (D. C. Eat.) Maxon. Rare; the type locality is in upper Teton Canyon, Teton Co.

PTERIDIUM Bracken Fern

Large, mesophytic ferns from deep-seated rhizomes, covered with hairs. Lvs deciduous, pinnately compound several times; petiole coarse, firm, and stem-like; pinnule margins revolute and covering the sori which are continuous along the margins.

1. **P. aquilinum** (L.) Kuhn. var. **pubescens** Underw. Frequent around the shores of some valley lakes such as Bradley and Leigh Lakes.

WOODSIA Woodsia Fern

Xerophytic ferns of rocky habitats, from short rhizomes, covered with brown scales; roots fibrous, brown. Lvs 2-3 times pinnate; sori on the veins of the pinnules, round; indusium arising from below the sorus and covering it when young, splitting at maturity into star-like segments, often inconspicuous in older sori.

1 Frond blades and petioles glabrous or somewhat glandular .... **W. oregana**
1 Frond blades and petioles with glandless septate hairs and glandular .... **W. scopulina**

1. **W. oregana** D. C. Eat. Infrequent in rocks of the canyons.
2. **W. scopulina** D. C. Eat. Frequent in talus slopes up to 8,000 ft., major canyons.

CUPRESSACEAE Cypress Family

Monoecious or dioecious evergreen aromatic trees or shrubs; leaves opposite or whorled, scale-like or linear; staminate cones small; ovuliferous scales paired or whorled, (cones fleshy in *Juniperus*, resembling a berry).

JUNIPERUS Juniper

With the characteristics of the family

1 Leaves all needle-like to awl-like; jointed at base, in whorls of 3; plants usually prostrates shrubs .... **J. communis**

1 Leaves scale-like or, if needle-like, then decurrent on the twig, opposite or in whorls of 3; plants small trees .... **J. scopulorum**

1. **J. communis** L. var. **depressa** Pursh. Mountain Juniper. Frequent on the shores of the valley lakes and extending into some canyons.
2. **J. scopulorum** Sarg. Rocky Mountain Juniper. Infrequent on Blacktail Butte and in the Gros Ventre river bottoms.

PINACEAE Pine Family

Monoecious evergreen trees or shrubs with resinous wood and foliage; leaves needle-like, spirally arranged microsporophylls; ovulate cones large and woody, each ovuliferous scale bearing 2 ovules on the upper surface.

1 Leaves in clusters of 2-5; ovulate cones maturing the second season .... **Pinus**

1 Leaves solitary, not in clusters; ovulate cones maturing the first season.

  2 Leaves sharp and more or less square in cross section .... **Picea**

  2 Leaves blunt and flat in cross section

    3 Ovulate cones hanging downward, bearing conspicuous 3 pronged bracts between ovuliferous scales .. .. **Pseudotsuga**

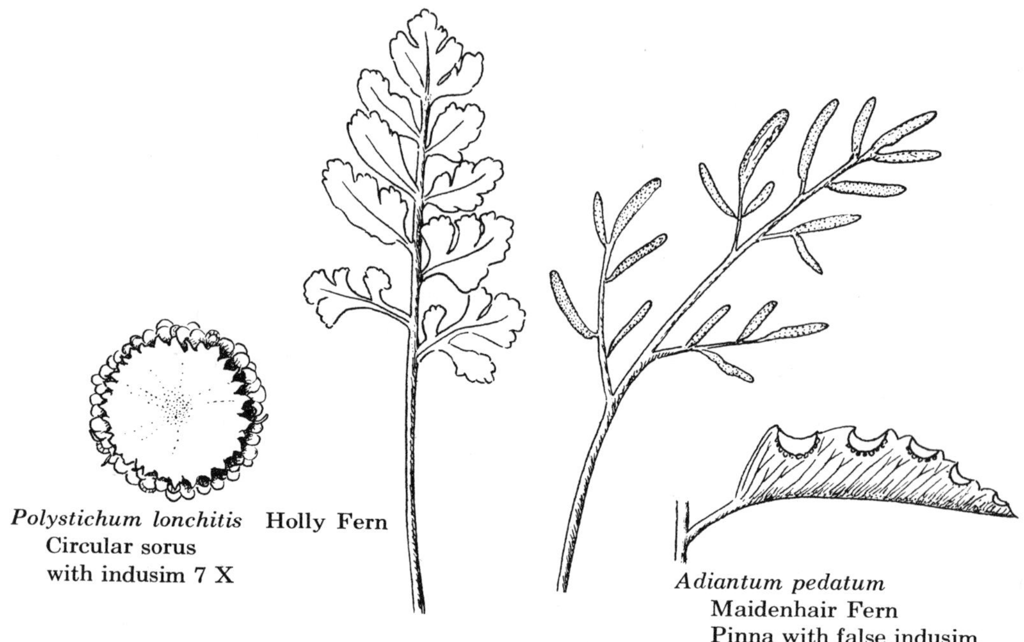

Polystichum lonchitis Holly Fern
Circular sorus
with indusim 7 X

Adiantum pedatum
Maidenhair Fern
Pinna with false indusim

Cryptograma crispa
Rock-brake Fern
right, fertile frond and left, sterile frond

Polypodiaceae

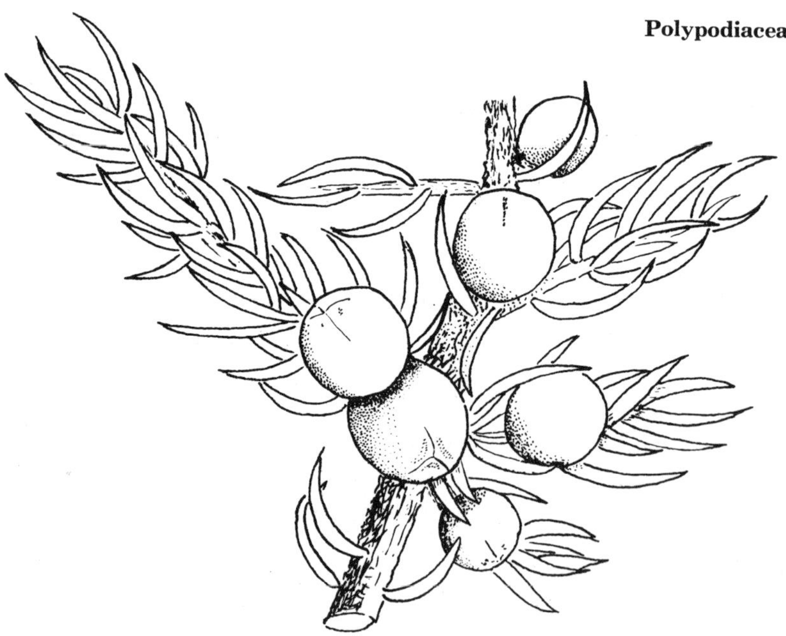

Juniperus communis
Common Juniper

Cupressaceae

3 Ovulate cones erect, ovuliferous scales deciduous at maturity .... **Abies**

ABIES Fir

Evergreen trees with a conical to spire-like shape; bark thin and smooth except for numerous horizontal resin pockets. Lvs flat, sessile, spirally arranged. Cones borne at the tip of previous year's branches; staminate cones 7-20 mm long, pendent; ovulate cones sessile and erect, mostly near the top of tree, both scales and bracts deciduous at maturity leaving a persistent peg-like axis.

1. **A. lasiocarpa** (Hook.) Nutt. Subalpine Fir, Alpine Fir. Abundant on moraines and mountains up to timberline.

PICEA Spruce

Evergreen trees; bark thin and scaly. Lvs spirally arranged, linear, sharp-pointed, 4-sided, and upon falling, leaving branches roughened with leaf bases. Cones produced on year-old twigs, the staminate cones 1-2 cm long, pendent; the ovulate cones, mostly pendent and deciduous as a unit, non-prickly scales; seeds winged.

1 Young branches usually pubescent; cones 3-5 cm long; leaves not rigid .... **P. engelmannii**
1 Young branches glabrous; cones 5-9 cm long; leaves stiff and pungent .... **P. pungens**

1. **P. engelmannii** Parry ex Engelm. Engelmann Spruce. Common on the west shores of the valley lakes and throughout the canyons up to about 8,500 ft.
2. **P. pungens** Engelm. Abundant along the Snake River.

PINUS Pine

Evergreen trees with scaly or furrowed bark. Lvs of 2 kinds, some needle-like and green, borne in clusters of 2-5 (in ours) on spur branches, the others scale-like and membranous. Staminate cones numerous, falling soon after pollen loss; ovulate cones single to clustered, maturing in 2-3 years, ovuliferous scales persistent; seeds winged.

1 Leaves 2 in a cluster .... **P. contorta**
1 Leaves 5 in a cluster
2 Ovulate cones 2.5-7 cm long, purple .... **P. albicaulis**

2 Ovulate cones 7-12 cm long, green or brown ...... **P. flexilis**

1. **P. albicaulis** Engelm. Whitebark Pine. Common in the canyons above 8,000 ft.
2. **P. contorta** Dougl. var. **latifolia** Engelm. Lodgepole Pine. Abundant throughout the valley, especially in pot holes and on moraines.
3. **P. flexilis** James. Limber Pine. Frequent on moraines of the valley.

PSEUDOTSUGA Douglas Fir

Evergreen trees with deeply furrowed bark. Lvs needle-like, spirally arranged, flattened, constricted at the base. Cones of both sexes borne on the twigs of the previous years; staminate cones solitary in axils of the leaves; ovulate cones pendent, maturing in one season, ovuliferous scales subtended and exceeded by a 3-lobed bract; seeds broadly winged.

1. **P. menziesii** (Mirb.) Franco. Common on the moraines and in the canyons below 8,500 ft.

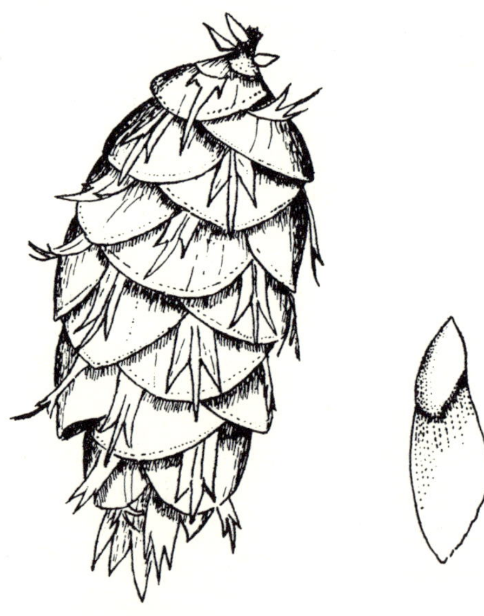

Pseudotsuga menziesii
Douglas Fir
Seed

Pinaceae

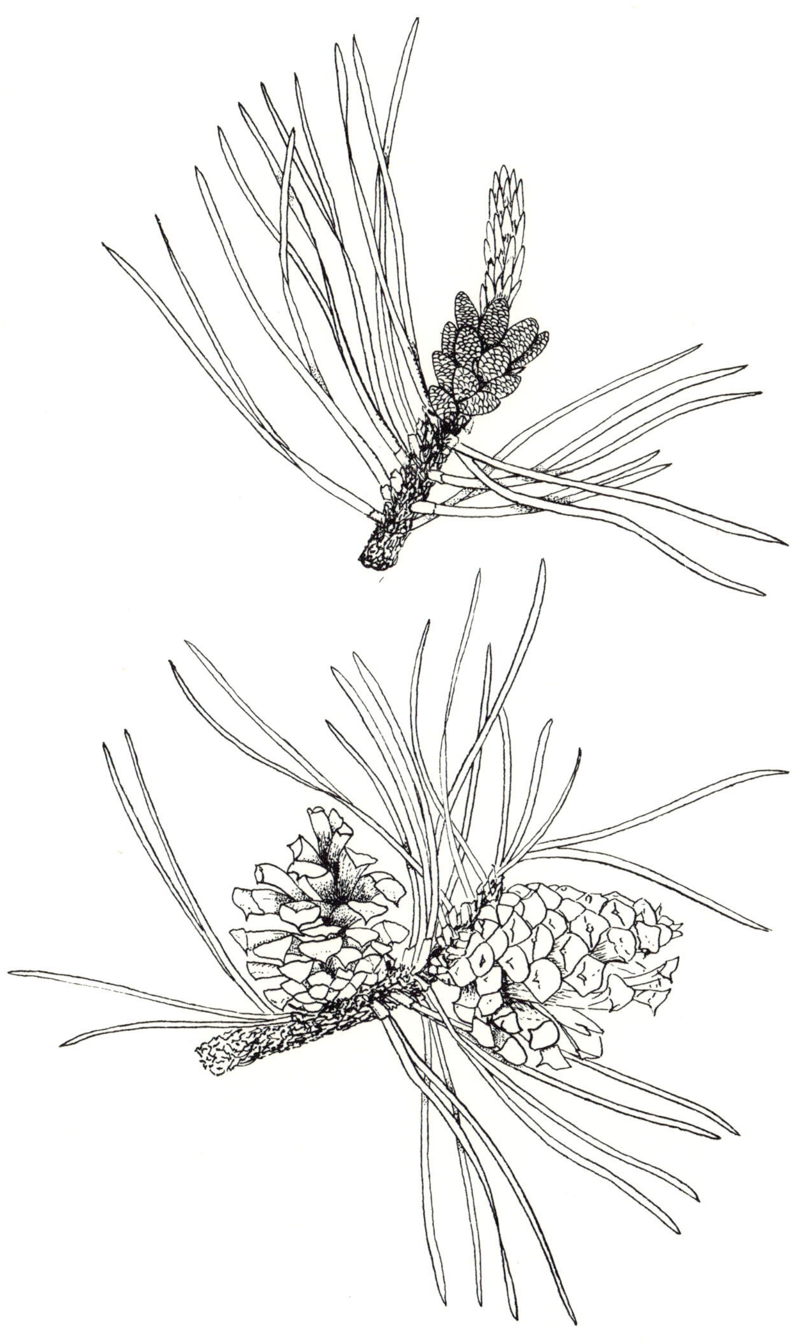

Pinus contorta
Lodgepole Pine

Pinaceae

Key to the Flowering Plants

CLASS MAGNOLIOPSIDA (DICOTS)

Key to the Groups

1 Perianth none or of a single whorl, the parts all much alike in color, size and texture, (no corolla present) . GROUP 1, P. 23
1 Perianth composed of two sets of parts (sepals and petals)
  2 Corolla of separate petals (no union)
    3 Stamens numerous, more than twice as many as the petals .... GROUP 2, P. 25
    3 Stamens usually less than 10, not more than twice as many as the petals .... GROUP 3, P. 26
  2 Corolla of united petals (union at least near the base) .... GROUP 4, P. 28

GROUP 1. *Perianth None or of a Single Whorl*

1 Plants parasitic on the branches of woody plants .... LORANTHACEAE, P. 135
1 Plants not parasitic on branches, roots anchored in soil
  2 Plants vines
    3 Leaves compound; stamens and pistils many .... RANUNCULACEAE *(Clematis)*, P. 164
    3 Leaves simple, palmately lobed, stamens and pistils 5 or fewer .... MORACEAE *(Humulus)*, P. 139
  2 Plants trees, shrubs or herbs
    4 Flowers (at least staminate ones) in catkins or catkin-like spikes
      5 Plants dioecious; a single flower to each scale of the catkin; seeds with a tuft of fine hairs .... SALICACEAE, P. 183
      5 Plants monoecious; 2-3 flowers to each scale; seeds lack hairs .... BETULACEAE, P. 73
    4 Flowers neither in catkins nor in catkin-like spikes
      6 Plants aquatic, usually more or less submerged
        7 Ovary inferior .... HALORAGACEAE, P. 127
        7 Ovary superior
          8 Leaves entire, opposite, often crowded into terminal rosettes ...... CALLITRICHACEAE, P. 90
          8 Leaves largely dissected, whorled .... CERATOPHYLLACEAE, P. 101
      6 Plants terrestrial, sometimes growing in wet soil
        9 Plants shrubs or subshrubs

10 Plants spinose, fruit a berry .... BERBERIDACEAE, P. 72
10 Plants unarmed, fruit not as above
11 Ovary inferior or appearing so
12 Ovary 1-loculed, stellate pubescence .... CORNACEAE, P. 105
12 Ovary 2-loculed, no stellate pubescence .... ELAEAGNACEAE, P. 108
11 Ovary superior
13 Stamens 2-9
14 Flowers subtended by an involucre of more or less united bracts; achenes lenticular or triangular .... POLYGONACEAE, P. 151
14 Flowers without an involucre of bracts; fruit usually a utricle .... CHENOPODIACEAE, P. 101
13 Stamens 15 or more .... ROSACEAE *(Cercocarpus)*, P. 172
9 Plants herbaceous
15 Plants with stinging hairs .. URTICACEAE, P. 207
15 Plants with various pubescence, but not as above
16 Ovary superior
17 Flowers perfect; stamens 6-9; fruit an achene .... POLYGONACEAE, P. 151
17 Flowers perfect or unisexual; stamens 1-5; fruit an utricle or capsule
18 Sepals and bracts more or less dry, thin and nongreen .... AMARANTHACEAE, P. 33
18 Sepals and bracts green and herbaceous ...... CHENOPODIACEAE, P. 101
16 Ovary inferior
19 Ovary 2-loculed, 1 ovule in each locule; fruit 2-seeded
20 Perianth united, leaves opposite or whorled, inflorescence usually a cyme .... RUBIACEAE, P. 182
20 Perianth of separate segments; leaves alternate or basal; inflorescence is an umbel .... APIACEAE, P. 33

19 Ovary 1-loculed, sometimes involving several carpels but with a single ovule developing; fruit 1-seeded
21 Flowers without pedicels, in head type of inflorescence; anthers united into a tube ASTERACEAE, P. 41
21 Flowers not as above; anthers distinct
22 Leaves alternate .... SANTALACEAE, P. 192
22 Leaves opposite or basal .... VALERIANIACEAE, P. 208

GROUP 2. *Corolla of Separate Petals, Stamens More Than Twice As Many As the Petals*

1 Ovary inferior
2 Stems thick and succulent, spiny; petals numerous; leaves usually absent .... CACTACEAE, P. 90
2 Stems not thick and succulent, sometimes spiny; petals less than 10; leaves present
3 Plants herbaceous, covered with rough hairs .... LOASACEAE, P. 135
3 Plants shrubs or small trees; rough hairs absent
4 Leaves alternate; fruit a pome ...... ROSACEAE, P. 172
4 Leaves opposite; fruit a capsule SAXIFRAGACEAE, P. 193
1 Ovary superior
5 Plants aquatic, with broad floating leaves .... NYMPHAEACEAE, P. 139
5 Plants terrestrial, but often growing in wet places, or if in water the leaves smaller and dissected
6 Filaments united into a tube around the pistil .... MALVACEAE, P. 137
6 Filaments not united into a tube, distinct
7 Leaf margins dotted with colored depressions .. .. HYPERICACEAE, P. 139
7 Leaf margins not dotted with colored depressions
8 Sepals 2, pistils only 1 in each flower .... PORTULACACEAE, P. 155
8 Sepals 3-5, pistils more than 1 in each flower

9 Stamens inserted on a fleshy disc .... PAEONIACEAE, P. 145
9 Stamens not as above, fleshy disc absent .. .. RANUNCULACEAE, P. 164

GROUP 3. *Corolla of Separate Petals, Stamens Few, Not More Than Twice As Many As the Petals*

1 Ovary inferior
2 Plants aquatic .... HALORAGACEAE, P. 127
2 Plants terrestrial
3 Ovules and seeds more than one in each locule
4 Ovary 1-loculed; fruit a berry . SAXIFRAGACEAE, P. 193
4 Ovary 2 to many locules
5 Plants shrubs, leaves evergreen .... CELASTRACEAE, P. 100
5 Plants herbs
6 Style 1; stamens 4-8 ...... ONAGRACEAE, P. 139
6 Style 2-3; stamens 5 or 10 .... SAXIFRAGACEAE, P. 193
3 Ovules and seeds only 1 in each locule
7 Stamens 4-8
8 Plants shrubs; fruit a drupe .... CORNACEAE, P. 105
8 Plants herbs; fruit a capsule .. ONAGRACEAE, P. 139
7 Stamens 5 or 10
9 Stamens 10; plants shrubs ...... ROSACEAE, P. 172
9 Stamens 5; plants herbs or shrubs
10 Plants herbs, with mainly compound leaves .... APIACEAE, P. 33
10 Plants shrubs, with simple leaves .... RHAMNACEAE, P. 171
1 Ovary superior
11 Corolla irregular
12 Ovary simple .... FABACEAE, P. 112
12 Ovary compound, but 1-loculed
13 Petals 4; stamens 6 .... FUMARIACEAE, P. 120
13 Petals 5; stamens 5 .... VIOLACEAE, P. 210
11 Corolla regular
14 Flowers with more than 1 pistil
15 Plants succulent .... CRASSULACEAE, P. 106
15 Plants not succulent

16 Flowers perigynous; stamens inserted on a hypanthium . . . . ROSACEAE, P. 172

16 Flowers hypogynous; stamens inserted below pistils . . . . RANUNCULACEAE, P. 164

14 Flowers with a single pistil

17 Plants aquatic or at least occupying marsh or bog areas

18 Plants with prominent grandular hairs for trapping insects . . . . DROSERACEAE, P. 108

18 Plants without prominent glandular hairs

19 Plants from stout creeping rootstocks; sheathing petioles . . . . MENYANTHACEAE, P. 137

19 Plants slender annuals (in ours); petioles not sheathing . . . . LIMNANTHACEAE, P. 134

17 Plants terrestrial, usually occupying drier sites

20 Plants trees or shrubs

21 Anthers opening by pores; leaves with spiny teeth . . . . BERBERIDACEAE, P. 72

21 Anthers not opening by pores; leaves without spiny teeth

22 Leaves opposite; fruits winged . . . . ACERACEAE, P. 32

22 Leaves alternate; fruit a drupe . . . . ANACARDIACEAE, P. 33

20 Plants herbs, or woody at the base only

23 Sepals and petals 4; stamens 6 (4 plus 2) . . . . BRASSICACEAE, P. 79

23 Sepals, petals and stamens variable, but not as above

24 Fruit a schizocarp (splitting at maturity into 5 parts with long beaks) . . . . GERANIACEAE, P. 123

24 Fruit a capsule

25 Ovary 10-loculed; fragile petals easily knocked off . . . . LINACEAE, P. 134

25 Ovary less than 10-loculed; petals firmly attached

26 Leaves compound, 3 leaflets . . . . OXALIDACEAE, P. 145

26 Leaves simple

27 Leaves evergreen or reduced to non-green scales PYROLACEAE, P. 159
27 Leaves not as above
28 Leaves opposite with distinctly swollen nodes; sepals 4-5 .... CARYOPHYLLACEAE, P. 94
28 Leaves opposite or alternate, nodes not swollen, sepals 2 (4-9 *Lewisia*) .... PORTULACACEAE, P. 155

GROUP 4. *Corolla of United Petals*

1 Ovary inferior or partly so
2 Stamens more than 5; pollen released by means of terminal pores .... ERICACEAE, P. 109
2 Stamens 5 or less; pollen not released as above
3 Plants shrubs or trailing subshrubs
4 Flowers in involucrate heads; ovary 1-loculed .... ASTERACEAE, P. 41
4 Flowers not in involucrate heads; ovary 1 to 5-loculed
5 Leaves opposite, 3 to 5-loculed .... CAPRIFOLIACEAE, P. 91
5 Leaves alternate; 1-loculed. GROSSULARIACEAE, P. 125
3 Plants herbaceous
6 Stamens 1-3, always fewer than the corolla lobes .... VALERIANACEAE, P. 208
6 Stamens usually 4-5, always of the same number as the corolla lobes
7 Flowers in involucrate heads .. ASTERACEAE, P. 41
7 Flowers not in involucrate heads
8 Corolla (in ours) blue or purple .... CAMPANULACEAE, P. 90
8 Corolla (in ours) pink or white .... RUBIACEAE, P. 182
1 Ovary superior
9 Plants nongreen, saprophytic or parasitic; leaves reduced to scales
10 Corolla bilabiate; stamens 4; plants root parasites .... OROBANCHACEAE, P. 144

10 Corolla not bilabiate; stamens 6-10; plants saprophytes . . . . PYROLACEAE, P. 159
9 Plants chlorophyllous, usually with greenish leaves
11 Plants with milky juice . . . . APOCYNACEAE, P. 41
11 Plants without milky juice
12 Corolla irregular
13 Plants aquatic; leaves often dissected . . . . LENTIBULARIACEAE, P. 134
13 Plants terrestrial, if aquatic then not as above
14 Stamens 10; petals 5; fruit a legume . . . . FABACEAE, P. 112
14 Stamens 2-6; petals 4-5; fruit various but not a legume
15 Sepals 2; stamens 6 . . . . FUMARIACEAE, P. 120
15 Sepals 5; stamens 2-5
16 Ovary unlobed or bilobed; ovules and seeds numerous in each locule . . . . SCROPHULARIACEAE, P. 197
16 Ovary becoming deeply 4-lobed at maturity, only one ovule or seed per lobe
17 Corolla only slightly irregular; stamens 4-5; style arising from apex of ovary . . . . VERBENACEAE, P. 208
17 Corolla strongly irregular; stamens 2 or 4; style arising from between the lobes of ovary . . LAMIACEAE, P. 131
12 Corolla regular
18 Stamens 10-many, more or less connate
19 Filaments united well above base; stamens many; leaves simple . . . . MALVACEAE, P. 137
19 Filaments united at base only; stamens 10; leaves compound . . . . OXALIDACEAE, P. 145
18 Stamens 4-10, not connate
20 Stamens free of corolla
21 Pistils more than 1; plants succulent . . . . CRASSULACEAE, P. 106
21 Pistils 1 per flower; plants not succulent
22 Anthers opening by a terminal pore . . . . ERICACEAE, P. 109

22 Anthers opening by a longitudinal slit .... PYROLACEAE, P. 159

20 Stamens epipetalous

23 Stamens opposite the corolla lobes; free central placentation ... PRIMULACEAE, P. 158

23 Stamens alternate with the corolla lobes; placentation various

24 Corolla veinless, scarious; top of capsule coming off as a lid .... PLANTAGINACEAE, P. 145

24 Corolla with veins, not scarious; capsule never as above

25 Ovary 1-loculed

26 Leaves opposite or whorled, entire; plants mostly glabrous ... . GENTIANACEAE, P. 122

26 Leaves usually alternate, if opposite not entire; plants mostly hairy .... HYDROPHYLLACEAE, P. 128

25 Ovary more than 1-loculed

27 Ovary deeply 4-lobed .... BORAGINACEAE, P. 73

27 Ovary not deeply 4-lobed

28 Ovary 3-loculed .... POLEMONIACEAE, P. 146

28 Ovary 2-loculed

29 Plants trailing or twining ...... CONVOVULACEAE, P. 105

29 Plants not trailing or twining

30 Style 1, stigma entire or 2-lobed; fruit a capsule or berry .... SOLANACEAE, P. 207

30 Styles 2 or 2-branched before reaching the stigmas; fruit a capsule . HYDROPHYLLACEAE, P. 128

CLASS LILIOPSIDA (MONOCOTS)

Key to the Families

1 Plants floating, without an apparent stem . . LEMNACEAE, P. 227
1 Plants with obvious stems and leaves, terrestrial or aquatic but not free floating
  2 Perianth lacking or reduced and inconspicuous, perianth often bristle-like bracts or scales, not petal-like in texture or color
    3 Plants not grasslike; flowers mostly imperfect, staminate and pistillate flowers are in heads or in elongate terminal spikes
      4 Flowers very numerous in a double spike, staminate above and pistillate below; plants usually over 10 dm tall . . . . TYPHACEAE, P. 272
      4 Flowers several in globose-capitate or axillary clusters; plants usually under 10 dm long
        5 Flowers in globose-capitate clusters; leaves alternate . . . . SPARGANIACEAE, P. 272
        5 Flowers in axillary clusters; leaves opposite . . . . ZANNICHELLIACEAE, P. 274
    3 Plants grasses or grasslike; flowers mostly perfect, if imperfect then usually borne in small spikes
      6 Leaf sheaths usually split lengthwise on the side opposite the blade; leaves in 2 rows; stems mostly hollow and round in cross-section; flowers subtended by 2 bracts . . . . POACEAE, P. 239
      6 Leaf sheaths not split; leaves in 3 rows; stems mostly solid and triangular in cross-section; flowers subtended by a single bract . . . . CYPERACEAE, P. 212
  2 Perianth conspicuous, the segments either separated into a calyx and corolla or all alike
    7 Plants submersed or floating aquatics, only the inflorescence above the surface; flower parts in 4's . . . . POTAMOGETONACEAE, P. 269
    7 Plants mostly terrestrial, if aquatic then the flowers other than 4-merous
      8 Pistils usually more than 6, 1-loculed and 1-ovuled, maturing into a cluster of achenes . . . . ALISMATACEAE, P. 212

8 Pistil 1, compound, 1, 3 or 6-loculed, maturing into a capsule, berry or follicle
9 Perianth mostly greenish to brownish or purplish green, all segments similar
10 Flowers in racemes or spikes; fruit follicular .... JUNCAGINACEAE, P. 226
10 Flowers in cymes or panicles; fruit a capsule .... JUNCACEAE, P. 223
9 Perianth usually more or less showy, the corolla often distinct in color and size from the calyx
11 Ovary superior (partially inferior in *Zigadenus elegans*) .... LILIACEAE, P. 227
11 Ovary inferior
12 Plants aquatic, usually submersed; mostly dioecious ...... HYDROCHARITACEAE, P. 222
12 Plants terrestrial; flowers perfect
13 Flowers regular; stamens 3 and free of style .... IRIDACEAE, P. 222
13 Flowers irregular; stamens 1 or 2 united with style .... ORCHIDACEAE, P. 234

CLASS MAGNOLIOPSIDA (DICOTYLEDONS)

ACERACEAE Maple Family

Woody. Lvs usually opp, simple/compound, no stip. Fls clustered, uni/bisex, reg, disk present, ov superior. Ca 4-5, Co 0-4-5, S 4-5, P2*. Samara.

ACER Maple

Characteristics of the family.

1 Flowers corymbose; petals present; a small shrub in our area, 1-2 m .... **A. glabrum**
1 Flowers umbellate; petals lacking; a small tree, 3-6 m.... .... **A. grandidentatum**

1. **A. glabrum** Torr. Rocky Mt. Maple. Infrequent on the lower slopes of the mountains and into the canyons at north end of the range.
2. **A. grandidentatum** Nutt. Bigtooth Maple. Frequent in lower portion of Snake River Canyon.

AMARANTHACEAE Amaranth Family

Herbs/shrubs. Lvs alt/opp, simple, no stip. Infl often racemose. Fls usually bisex, reg, ov superior. Per 3-5, scarious, S 3-5, P 2-3*. Circumscissle capsule (in ours) or utricle.

AMARANTHUS Amaranth; Pigweed

Characteristics of the family.

1 Flowers borne in terminal and axillary clusters; plants erect 5-20 dm; leaf blade longer than 3 cm .... **A. powellii**
1 Flowers borne in small axillary clusters; plants prostrate to erect up to 10 dm; leaf blade rarely longer than 3 cm . **A. albus**

†**1. A. albus** L. Widespread weed of disturbed sites.
†**2. A. powellii** Wats. Weed of disturbed sites.

ANACARDIACEAE Cashew Family

Woody, resinous. Lvs usually alt, simple/compound, no stip. Infl panicle. Fls usually bisex and reg, ov superior. Ca 3-5*, Co 0-5, S 5-10, P 3*. Drupe.

RHUS Sumac

Shrubs or woody vines. Lvs alt, 3-many pinnately arranged leaflets, often brightly colored in the fall. Fls usually perfect, but sometimes imperfect; 5 fertile stamens; ovary 1-loculed and 1-seeded.

1 Petals covered with soft spreading hairs on inner surface; fruit pubescent; 3-5 leaflets .... **R. trilobata**
1 Petals glabrous; fruit glabrous; 3 leaflets .... **R. radicans**

1. R. radicans L. Poison Ivy. Locally frequent on rocky outcropping 150 yds. south of hotspring area, west side of Jackson Lake.
2. R. trilobata Nutt. Squawbush. Not seen in G.T.N.P., but reported in Y.N.P.

APIACEAE (UMBELLIFERAE) Carrot Family

Herbs. Lvs alt often pinnately dissected, petioles sheathing. Infl umbel. Fls bisex, reg, ov inferior. Ca 5, often reduced, Co 5, S 5, P 2*. Schizocarp.

1 Leaves all simple and entire; plants of alpine areas . **Bupleurum**
1 Leaves mostly compound or deeply cleft
  2 Leaves mostly with well defined leaflets, not dissected into small and narrow portions
    3 Leaflets 3, very large, up to 4 dm wide and long . . . . **Heracleum**
    3 Leaflets usually more than 3, usually less than 1 dm wide
      4 Plants with fibrous or fleshy-thickened, fascicled roots; plants perennial
        5 Base of stem hollow with transverse partitions; some roots usually tuberous thickened **Cicuta**
        5 Base of stem lacking transverse partitions; roots not tuberous thickened . . . . **Sium**
      4 Plants with a taproot or stout caudex; plants annual, biennial or perennial
        6 Fruits almost round in cross-section or flattened laterally
          7 Fruits linear or linear oblong to club-shaped, not winged; leaves always with well defined leaflets . . . . **Osmorhiza**
          7 Fruits broader and often shorter, often some of the ribs winged; leaflets not well defined . . . . **Ligusticum**
        6 Fruits flattened dorsally

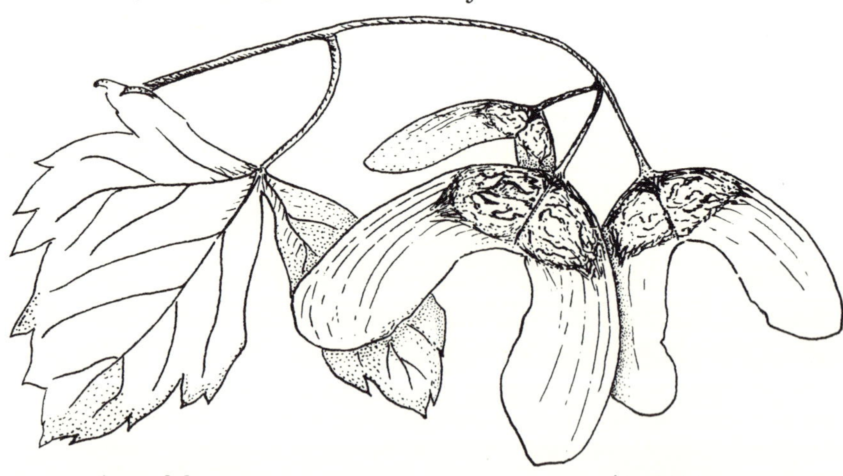

Acer glabrum
Rocky Mt. Maple

Aceraceae

8 Stylopodium well developed; petals white .... **Angelica**

8 Stylopodium scarcely developed; petals yellow .... **Lomatium**

2 Leaves mostly dissected into small and narrow segments, no well defined leaflets

9 Stems purple spotted; biennial weed of moist habitats; white flowers .... **Conium**

9 Stems lacking purple spots; habit and habitat variable

10 Plants with an elongated taproot; often with stout branching caudex .... **Cymopterus**

10 Plants with fibrous or fleshy-thickened roots

11 Plants with a single, globose, corm-like root; plants low, up to 1 dm tall .... **Orogenia**

Rhus radicans
Poison Ivy

Anacardiaceae

11 Plants with solitary or clustered thickened roots; plants taller, up to 12 dm tall . . . **Perideridia**

ANGELICA Angelica

Large perennials, usually with a single stem from a stout taproot. Lvs pinnately or ternately compound. Inflorescence usually of several compound umbels; calyx teeth usually lacking; petals usually white; fruit elliptic-oblong to orbicular, strongly flattened dorsally.

1 Leaves oblong to elliptic, pinnate; ovaries obscurely scabrous . . . . **A. pinnata**
1 Leaves more deltoid, ternate-pinnate; ovaries glabrous . . . . **A. arguta**

1. **A. arguta** Nutt. Frequent on streambanks and wet meadows of the valley.
2. **A. pinnata** Wats. Common in moist places on moraines of the valley.

BUPLEURUM

Leafy-stemmed annuals or perennials. Lvs simple and entire. Inflorescence of compound umbels; involucre of leafy bracts; calyx teeth usually lacking; petals yellow or sometimes purple; fruit oblong, 3-4 mm long, glaucous, the ribs are prominent.

1. **B. americanum** Coult. & Rose. Frequent east shore of Timberline Lake, 10,400 ft.

CICUTA Water Hemlock

Perennials with hollow stems and well developed transverse partitions at the stem base. Lvs one to three times pinnate or ternate-pinnate; basal and cauline. Inflorescence of compound umbels; flowers white or greenish; calyx teeth evident; fruit rotund-ovate to orbicular, 2-4 mm long.

†1. **C. douglasii** (DC.) Coult. & Rose. Not collected in the park, but expected soon. This plant is highly poisonous to man and livestock.

CONIUM Poison Hemlock

Glabrous biennials with purple-spotted stems and a stout taproot. Lvs pinnately or ternate-pinnately dissected with rather small, ultimate segments. Inflorescence of compound umbels; flowers white; calyx teeth lacking; fruit glabrous, flattened somewhat laterally, with wavy ribs.

†1. **C. maculatum** L. Roadside ditches and moist disturbed sites; an established poisonous weed from Eurasia.

CYMOPTERUS

Caulescent or acaulescent perennials with stout taproot. Lvs compound or decompound; petioles sheathing. Inflorescence of compound umbels; flowers white, yellow or purple; calyx teeth small or lacking; stylopodium lacking; fruit flattened dorsally, ribs usually winged.

1 Leaves bluish-green; mature pseudoscape 5-16 cm long . . . . **C. longipes**
1 Leaves green; pseudoscape lacking . . . . **C. hendersonii**

1. **C. hendersonii** (Coult. & Rose.) Cronq. [*Pteryxia hendersoni* Math. & Const.]. Frequent in major canyons above 9,000 ft.
2. **C. longipes** Wats. Teton Pass.

HERACLEUM Cow Parsnip

Coarse biennial or perennial herbs, 1-3 m tall, from a taproot or cluster of fibrous roots. Lvs large, ternately compound; petioles sheathing. In florescence of compound umbels; involucre lacking; flowers white; calyx teeth minute; the outer corolla lobes often bifid; fruit obovate to obcordate, strongly flattened dorsally.

1. **H. lanatum** Michx. Common in moist areas throughout the valley and lower parts of the canyons.

LIGUSTICUM Lovage

Perennial herbs from a taproot. Lvs petiolate, compound, petioles sheathing. Inflorescence of compound umbels; involucre wanting; flowers white or pinkish; calyx teeth evident or obscure; fruit only slightly flattened laterally, glabrous, ribs evident.

1. **L. filicinum** S. Wats. Frequent in moist aspen groves and coniferous forests of the valley and major canyons.

LOMATIUM Desert Parsley; Biscuitroot

Perennial herbs from a taproot. Lvs mostly basal, pinnately compound to dissected. Inflorescence of compound umbels; flowers mostly yellow or white, calyx teeth minute or lacking; fruit dorsally flattened with marginal wings. This is a difficult genus with many variable species.

1 Plants mostly 5-15 dm tall at maturity .... **L. dissectum**
1 Plants mostly 1-5 dm tall at maturity
  2 Leaves dissected into numerous small, usually crowded segments
    3 Plants usually glabrous; fruits glabrous or often granular roughened .... **L. cous**
    3 Plants sparsely to densely puberulent; fruits puberulent .... **L. foeniculaceum**
  2 Leaves with few divisions, leaflets linear to linear lanceolate
    4 Involucral bracts generally lacking .... **L. ambiguum**
    4 Involucral bracts present .... **L. triternatum**

1. **L. ambiguum** (Nutt.) Coult. & Rose. Frequent on roadside cuts north of Moose.
2. **L. cous** (Wats.) Coult. & Rose. [*L. montanum* Coult. & Rose.]. Frequent on the outwash plain east of Moose.
3. **L. dissectum** (Nutt.) Math. & Const. var. **multifidum** (Nutt.) Math. & Const. Frequent on valley moraines and the North Fork of Cascade Canyon.
4. **L. foeniculaceum** (Nutt.) Coult. & Rose. Frequent on hills above the Gros Ventre River, T42N, R115 W.
5. **L. triternatum** (Pursh) Coult. & Rose. ssp. **triternatum.** Frequent in area of the Oxbow Bend of the Snake River and Teton Canyon.

OROGENIA Indian Potato; Turkey Peas

Scapose perennial herbs from a tuberous root. Lvs compounded into groups of three, with elongate ultimate segments. Inflorescence a compound umbel; involucre lacking; calyx teeth lacking; flowers white; anthers and apex of ovary purplish; fruit oval, glabrous.

1. **O. linearifolia** Wats. Frequent in sagebrush in vicinity of Park Hdqts., blooming as soon as the ground is free of snow.

OSMORHIZA Sweet Cicely

Perennial herbs from thick roots. Lvs petiolate, ternately to pinnately compound. Loose compound umbels; involucre lacking;

Lomatium ambiguum
Desert parsley

Apiaceae

flowers white, yellow, purple, or pink; calyx teeth lacking; fruit narrow, clavate or linear, somewhat compressed laterally, bristly to glabrous.

1 Fruit glabrous; stems clustered; flowers yellow . **O. occidentalis**
1 Fruit with bristly hairs; stems mostly single; flowers other than yellow
  2 Fruit concavely narrowed to the apex, the terminal 2 mm set off as a beaklike portion .... **O. chilensis**
  2 Fruit convexly narrowed to the rounded apex, the terminal 2 mm not beaklike .... **O. depauperata**

1. **O. chilensis** Hook. & Arn. Frequent in shade of wooded areas of the valley.
2. **O. depauperata** Phil. Frequent in wooded areas of the valley.
3. **O. occidentalis** (Nutt.) Torr. Common in aspen stands in the vicinity of the Oxbow Bend.

PERIDERIDIA Yampah

Perennial herbs from tuberous, fascicled roots. Lvs mostly basal, pinnately compound or dissected. Inflorescence of compound umbels; fls white; calyx teeth obvious; fruit linear-oblong, compressed laterally, glabrous, with ribs.

1 Leaves dissected with many ultimate segments, segments dimorphic (some much longer than others); petioles dilated .... **P. bolanderi**
1 Leaves with only a few segments, not dimorphic; petioles not dilated .... **P. gairdneri**

1. **P. bolanderi** (Gray) Nels. & Macbr. Infrequent in sagebrush flats near Moose.
2. **P. gairdneri** (Hook. & Arn.) Mathias. Common in open woods throughout the valley. Roots are edible.

SIUM Waterparsnip

Perennial herbs from fibrous roots. Lvs pinnately compound with distinct toothed leaflets. Inflorescence of compound umbels; flowers white; calyx teeth lacking; fruits elliptic to orbicular, glabrous, somewhat compressed laterally.

1. **S. suave** Walt. Frequent in ponds of the northern part of the valley.

APOCYNACEAE Dogbane Family

Woody/herbs, with milky juice. Lvs opp entire, usually no stip. Infl racemose/cymose/fls solitary. Fls bisex, reg, ov superior, Ca 5*, Co 5*, S 5, P 2*, often united only by style. Fruit variable, seeds often with long hairs.

APOCYNUM Dogbane

Perennial herbs with milky juice. Lvs opp, nearly sessile to petiolate, entire. Fls borne in cymes; calyx 5 lobed; corolla tubular to campanulate, pink, the tube with appendages; stamens 5; fruit a pair of follicles, elongate; seeds with hair. A critical genus with considerable hybridization between species.

1. **A. androsaemifolium** L. var. **androsaemifolium.** Frequent on dry soils of the moraines.

ASTERACEAE (COMPOSITAE) Sunflower Family

Herbs/woody, sometimes with milky juice. Lvs alt/opp/basal, simple/dissected, no stip. Fls in involucrate heads. Fls uni/bisex, reg/irreg, ov inferior. Ca when present modified as a pappus, Co 5*, S 5*, P 2*. Achene.

1A Flowers all ligulate and perfect; milky juice
  2 Pappus of plumose bristles
    3 Achenes beaked at apex; involucral bracts usually more than 15 mm long; flowers yellow or purple . . . . **Tragopogon**
    3 Achenes not beaked, truncate at the apex; involucral bracts usually less than 15 mm long; flowers pink . . . . **Stephanomeria**
  2 Pappus of simple bristles, awns, or scales
    4 Pappus of 1-3 series of unawned or awned scales
      5 Pappus of a single series of awned scales . . **Microseris**
      5 Pappus of 2 or 3 series of unawned scales; corollas blue . . . . **Cichorium**
    4. Pappus of capillary bristles
      6 Achenes more or less flattened; stems leafy; heads in panicles or umbels
        7 Achenes beaked; flowers yellow or blue . . **Lactuca**

Apocynum androsaemifolium
Dogbane

Apocynaceae

7 Achenes truncate, not beaked; flowers yellow . . . . **Sonchus***

6 Achenes not flattened; stems leafy or scapose; heads solitary or variously placed

8 Leaves all basal; heads solitary on scapose peduncles

9 Achenes 4-5 ribbed, with minute spines at least near the apex . . . . **Taraxacum**

9 Achenes 10 ribbed or 10 nerved, without minute spines on the surface . . . . **Agoseris**

8 Leaves not all basal, the stems leafy; heads not on scapose pendunceles

10 Pappus white; involucral bracts thickened at base or on midrib . . . . **Crepis**

10 Pappus brown or reddish; involucral bracts not thickened . . . . **Hieracium**

1B Flowers, or some of them, tubular and eligulate; ligulate flowers if present are either pistillate or neutral, never perfect; no milky juice

Key to Groups of 1B

1 Heads with ray flowers

2 Ray flowers yellow or orange

3 Pappus chaffy, or of firm awns or absent; receptacle chaffy, bristly or naked . . . . Group 1

3 Pappus partly or entirely of numerous capillary, (sometimes plumose) bristles; receptacle naked . Group 2

2 Ray flowers white, pink, red, purple or blue, not yellow or orange . . . . Group 3

1 Heads with only disk flowers

4 Pappus partly or entirely of many capillary, sometimes plumose bristles . . . . Group 4

4 Pappus of awns or scales or very short, chaffy bristles or of a few low teeth, never plumose . . . . Group 5

Group 1. *Ray Flowers Orange or Yellow; Pappus Chaffy or Of Firm Awns or Absent*

1 Receptacle naked, or with a single row of chaff between the ray and disk flowers

*Genera not represented in the flora yet, but weedy spp. expected soon.

2 Plants annual; strong oily odor .... **Madia**
2 Plants biennial or perennial; odor not as above
3 Ray flowers short; usually 1-5 mm long ..... **Gutierrezia**
3 Ray flowers long; usually 5-30 mm long
4 Involucral bracts covered with sticky gum, the tips recurved .... **Grindelia**
4 Involucral bracts not covered with sticky gum
5 Involucral bracts few, mostly 5-13; achene elongate and narrow, 4 times as long as wide .... **Eriophyllum**
5 Involucral bracts numerous, mostly 20 or more; achenes short only 2-3 times as long as wide
6 Involucral bracts loose, eventually becoming reflexed .... **Helenium**
6 Involucral bracts appressed, always erect .. .. **Hymenoxys**
1 Receptacle chaffy or bristly throughout
7 Plants annual
8 Stems 1-10 dm tall; leaves mostly opposite .... **Bidens**
8 Stems 4-20 dm tall; leaves mostly alternate .. **Helianthus**
7 Plants perennial
9 Plants scapose, the cauline leaves, if any, reduced and inconspicuous .... **Balsamorhiza**
9 Plants obviously leafy stemmed
10 Cauline leaves, or at least the lower ones, opposite
11 Pappus absent; ray flowers mostly 7-17 mm long .... **Viguiera**
11 Pappus present and persistent; ray flowers mostly 1.5-5 cm long .... **Helianthella**
10 Cauline leaves all alternate
12 Receptacle flat or slightly convex (remove flowers to see this); leaves entire or slightly toothed .... **Wyethia**
12 Receptacle conic or columnar; leaves various
13 Ray flowers subtended by receptacular chaffy bracts; fruits flattened .... **Ratibida**
13 Ray flowers not subtended by receptacular chaffy bracts; fruits quadrangular ... **Rudbeckia**

Group 2. *Ray Flowers Orange or Yellow; Pappus Partly or Entirely of Capillary Bristles*

1 Leaves mostly opposite .... **Arnica**
1 Leaves alternate or basal
  2 Involucral bracts in one series, equal in size, narrow, except for a few short ones at the base of the head .. **Senecio**
  2 Involucral bracts in 2 or more series, equal in size or imbricate
    3 Plants without a taproot, having numerous fibrous roots from a rhizome or short caudex; heads generally numerous and small; pappus bristles mostly equal and white .... **Solidago**
    3 Plants usually with a taproot; heads larger; pappus various
      4 Pappus simple, bristles generally unequal, but not distinctively divided into two lengths .... **Haplopappus**
      4 Pappus double, the outer portion inconspicuous and shorter than the inner portion
        5 Leaves linear or nearly so, 0.5-3 mm wide; style appendages very short 0.5 mm long or less .... **Erigeron**
        5 Leaves broader, mostly over 3 mm wide; style appendages longer, mostly 0.7 mm long or more .... **Chrysopsis**

Group 3. *Ray Flowers White, Pink, Red, Purple or Blue*

1 Receptacle chaffy or bristly throughout the head; ray flowers white or pink
  2 Leaves entire .... **Wyethia**
  2 Leaves finely dissected .... **Achillea**
1 Receptacle naked or with a row of chaff between the ray flowers and disk flowers; ray flower color various
  3 Pappus of the disk flowers of scales, flattened bristles or a small crown
    4 Pappus of a short crown .... **Chrysanthemum**
    4 Pappus of about 10 or more flattened, bristle-like scales .... **Townsendia**
  3 Pappus of the disk flowers composed mostly of capillary bristles

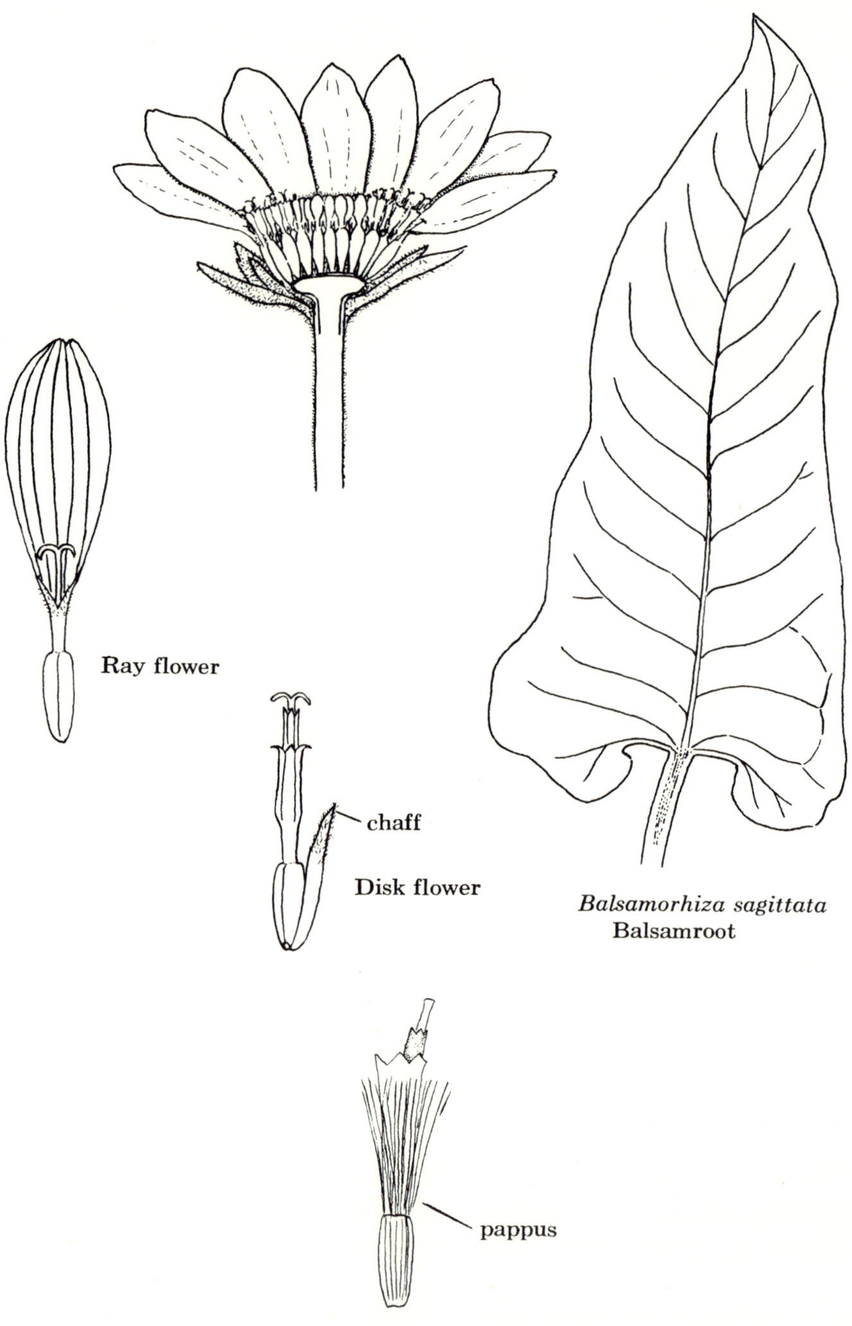

Senecio integerrimus Groundsel

Asteraceae

5 Involucral bracts about equal or more or less imbricate, neither leafy nor with papery base and herbaceous green tip; style appendages 0.5 mm long or less .... **Erigeron**

5 Involucral bracts either about equal and the outer bracts leafy, or usually imbricate, with papery base and green tips; style appendages usually more than 0.5 mm long

6 Plants annual or biennial or perennial from a distinct taproot; involucral bracts papery or coriaceous toward the base, often spreading at herbaceous tip .... **Machaeranthera**

6 Plants mostly perennial and if so, rhizomatous or fibrous rooted; involucral bracts mostly herbaceous .... **Aster**

Group 4. *Heads Composed of Disk Flowers Only; Pappus Capillary*

1 Plants with spiny-margined leaves and usually also with spiny involucre

2 Pappus bristles plumose .... **Cirsium**

2 Pappus bristles only barbellate .... **Carduus**

1 Plants without spiny leaves or involucre

3 Plants shrubby

4 Involucral bracts 4-6, equal .... **Tetradymia**

4 Involucral bracts more numerous, imbricate .... **Chrysothamnus**

3 Plants herbaceous

5 Leaves opposite; flowers yellow or orange .... **Arnica**

5 Leaves alternate, or if rarely opposite or whorled, then the flowers not yellow or orange

6 Plants more or less white-woolly; involucral bracts with dry scarious, tips

7 Taprooted annuals or perennials .... **Gnaphalium**

7 Fibrous-rooted perennials, often with rhizomes or stolons, but without a taproot

8 Basal leaves forming a conspicuous, persistent tuft; stems seldom very leafy .... **Antennaria**

8 Basal leaves soon deciduous, not much larger than the numerous and well developed cauline leaves .... **Anaphalis**

6 Plants not at all white-woolly, though often otherwise pubescent; involucral bracts not markedly scarious at the tip

9 Plants annual .... **Conyza**

9 Plants perennial

10 Flowers white to creamy .... **Brickellia**

10 Flowers yellow or orange .... **Senecio**

GROUP 5. *Heads Composed of Disk Flowers Only; Pappus Chaffy or of Awns or Absent.*

1 Plants annual

2 Leaves 1-3 times pinnatifid; plants 5-40 cm tall . **Matricaria**

2 Leaves ovate with long petioles; plants up to 20 dm tall .... **Iva**

1 Plants biennial or perennial

3 Receptacle elongate; large black heads .... **Rudbeckia**

3 Receptacle flat or merely convex; heads not black

4 Plants with an evident, though sometimes short, pappus of distinct awns or scales .... **Chaenactis**

4 Plants without a pappus, or the pappus a small minute crown

5 Heads greenish-yellow, in spikes, racemes or panicles .... **Artemisia**

5 Heads yellow, in flat topped clusters ...... **Tanacetum**

ACHILLEA Yarrow

Strong-scented herbs, perennial from rhizomes. Lvs alt, pinnately dissected. Heads many, corymbose. Ray fls 4-5, pistillate, white or pinkish; disk fls yellow. Pappus none. Achenes flattened. Receptacle chaffy.

1. **A. millefolium** L. ssp. **lanulosa** (Nutt.) Piper. Common throughout the valley.
1. **A. millefolium** L. ssp. **lanulosa** (Nutt.) Piper var. **alpicola** (Rydb.) Garrett. Infrequent in tundra above 10,000 ft.

AGOSERIS False Dandelion

Herbs with milky juice. Lvs mostly basal. Ligulate fls yellow or orange. Involucral bracts imbricated. Achenes usually 10 ribbed or nerved, beaked or beakless. Pappus-bristles fine and numerous, capillary, white.

1 Achenes with a beak about half as long as the body, (1-4 mm long); flowers yellow, often turning pink on drying . . . . **A. glauca**
1 Achenes with a beak more than half as long as the body of the achene; flowers burnt orange, commonly turning purple in age . . . . **A. aurantiaca**

1. **A. aurantiaca** (Hook.) Greene. Frequent in meadows and open woods at moderate elevations in the mountains.
2. **A. glauca** (Pursh) Raf. Frequent in meadows at all elevations.

ANAPHALIS Pearl Everlasting

Perennial herbs with white-woolly pubescence. Lvs alt, entire, basal lvs early deciduous. Heads several, discoid, many flowered; involucral bracts imbricate in several series, whitish, almost dry and scarious; distinct capillary bristles. This genus may be confused with species of *Antennaria* and *Gnaphalium*; it differs from the former in its equably leafy stem and from the latter by having well developed creeping rhizomes.

1. **A. margaritacea** (L.) Benth. & Hook. Frequent at Colter Bay Village and open sites in Indian Paintbrush Canyon.

ANTENNARIA Pussytoes

Cottony perennial herbs. Lvs basal or alt, entire. Heads small, fls all tubular and whitish. Receptacle naked; pappus of capillary bristles; the bristles of the ♂ fls more or less thickened above. Achene oblong, terete or slightly compressed. Because of the presence of apomixis, this genus is very difficult and the specific lines are blurred.

1 Plants with a single flower head at each stem tip; flowering stems mostly less than 1 dm tall . . . . **A. dimorpha**
1 Plants with more than one flower head on each stem
  2 Upper surfaces of the basal leaves distinctly less pubescent than the lower becoming glabrate sooner or later . . . . **A. racemosa**

2 Upper surfaces of the basal leaves nearly as pubescent as the lower, becoming glabrate only in extreme age
3 Plants without stolons; plants with clustered stems arising from a branched caudex .... **A. luzuloides**
3 Plants forming mats with numerous leafy stolons
4 Terminal tip of the involucral bracts brownish to blackish green; plants of montane or alpine areas
5 Terminal tip of the involucral bract usually sharp-pointed; bracts blackish green throughout .... **A. alpina**
5 Terminal tip of the involucral bracts usually blunt; bracts becoming white or pale brown .... **A. umbrinella**
4 Terminal tip of the involucral bracts white or pink, sometimes with a basal dark spot; plants of the mountains, but never alpine
6 Involucral bracts with a conspicuous black spot at the base of the papery portion; basal leaves narrowly oblanceolate .... **A. corymbosa**
6 Involucral bracts seldom if ever black at the base of the papery portion; leaves usually broader
7 Involucral bracts 4-6 mm high; bracts (in ours) pink .... **A. microphylla**
7 Involucral bracts 7-9 mm high; bracts genrally bright white .... **A. parvifolia**

1. **A. alpina** (L.) Gaertn. var. **media** (Greene) Jeps. Frequent in subalpine cirques.
2. **A. corymbosa** E. Nels. Common in the sagebrush flats west of Moose.
3. **A. dimorpha** (Nutt.) T. & G. Infrequent in sagebrush flats of the valley.
4. **A. luzuloides** T. & G. Frequent on rocky flats of the valley and slopes of the Teton Range.
5. **A. microphylla** Rydb. Common in sagebrush flats and moraines of the valley. [*A. arida* E. Nels., *A. rosea* (Eat.) Greene].
6. **A. parvifolia** Nutt. Common on dry terraces of the Pacific Creek area. [*A. aprica* Greene].
7. **A. racemosa** Hook. Common on the north exposure of Signal Mt. and major canyons.

8. **A. umbrinella** Rybd. Frequent near outlet of Schoolroom Glacier Lake, 10,000 ft.

ARNICA Arnica

Perennial herbs. Lvs mostly opp, simple. Heads many-fls, usually large and radiate, solitary or corymbose. Fls yellow. Ray fls (if present) pistillate. Receptacle flat, naked or pubescent; pappus of capillary bristles. Achenes 5-10 ribbed. A critical genus difficult to separate into distinct species.

1 Stem have more than 4 pairs of leaves; petioles short; blades mostly lack teeth
  2 Single stem; involucral bracts with a tuft of long white hairs at or near the tip .... **A. chamissonis**
  2 Several stems; involucral bracts not markedly more hairy at the tip .... **A. longifolia**
1 Stems with no more than 4 pairs of leaves, most of them at or near the base; blades often with teeth
  3 Heads without ray flowers, the immature heads nodding .... **A. parryi**
  3 Heads with ray flowers, the immature heads erect
    4 Pappus subplumose, more or less yellowish-brown; no evident basal leaves .... **A. mollis**
    4 Pappus merely barbellate, usually white; basal leaves often conspicuous
      5 Plants hairy; leaves ovate, tending to be cordate, basal and stem leaves distinctly petiolate . **A. cordifolia**
      5 Plants sparingly hairy; basal leaves not cordate, but ovate-truncate .... **A. latifolia**

1. **A. chamissonis** Less. ssp. **foliosa** (Nutt.) Maguire. Locally common in moist places in open forest, Snake River bottoms.
2. **A. cordifolia** Hook. Common in the coniferous forests of the valley.
3. **A. latifolia** Bong. Frequent in the major canyons, 7,000-8,000 ft.
4. **A. longifolia** D. C. Eat. ssp. **longifolia.** Locally common in canyons and couloirs.
5. **A. mollis** Hook. var. **mollis.** Frequent in bog west of Taggart Lake.
6. **A. parryi** A. Gray. Frequent in Moose Ponds west of Moose.

ARTEMISIA Sagebrush; Wormwood

Contributed by Robert D. Dorn

Annual, biennial, or perennial herbs or shrubs, usually aromatic. Lvs alt, entire to dissected. Ray fls lacking. Involucral bracts usually imbricate, dry, often with scarious margins. Receptacle naked or long-hairy, rarely glandular. Pappus usually lacking.

1 Flowers all perfect; shrubs, rarely appearing like subshrubs
  2 Leaves linear, linear-oblanceolate, or linear-elliptic, entire or a few sometimes irregularly once or twice lobed or toothed .... **A. cana**
  2 Leaves, or many of them, 3 toothed or 3-6 parted at tip, often cuneate
    3 Leaves mostly deeply cleft into 3-6 linear divisions, the basal part of leaf usually about as wide as the divisions and not broadened .... **A. tripartita**
    3 Leaves mostly 3 toothed at tip, or if lobed, the basal part of at least some leaves usually obviously broadened below the lobes
      4 Plants mostly 1-4 dm high, rarely higher, often glandular-punctate; inflorescence spike-like or a narrow panicle mostly less than 1.5 cm wide, rarely wider; rarely above 8,000 ft ...... **A. arbuscula**
      4 Plants mostly over 4 dm high, rarely shorter, glands present or not; inflorescence usually a panicle often over 1.5 cm wide; sometimes above 8,000 ft.
        5 Heads moderate in number, somewhat loosely arranged, averaging 4-5 mm long and 3-4 mm wide; some leaves usually over 3 cm long and irregularly lobed at tip; usually well up in the mountains .... **A. rothrockii**
        5 Heads usually many and crowded, mostly smaller; leaves often shorter and usually regularly toothed at tip; lowlands to mountains .... **A. tridentata**

1 Flowers pistillate at margin of head, these often few and reduced, the middle ones perfect or sometimes the ovary aborted in flowers at very middle; herbs or subshrubs
  6 Flowers at middle of head fertile, the ovary normal

7 Receptacle with long hairs between the flowers
8 Heads numerous, mostly 5 mm or less wide; involucral bracts without brown or black margins; mostly low and middle elevations ..... **A. frigida**
8 Heads few, mostly 5-12 mm wide; involucral bracts with conspicuous brown or black margins; high elevations .... **A. scopulorum**
7 Receptacle not hairy
9 Plants annual or biennial with a taproot; leaves often glabrous or nearly so .... **A. biennis**
9 Plants perennial from a rhizome or caudex or rarely a taproot; leaves usually hairy at least beneath
10 Leaves primarily basal, the stem leaves few and progressively reduced upward; lower leaves usually 2-3 times cleft or divided; alpine or subalpine .... **A. norvegica**
10 Leaves mostly cauline, often less divided; mostly not alpine
11 Leaves finely 2-3 times pinnately divided, the rachis and divisions usually 1 mm or less wide; escaped cultivar .... **A. abrotanum**
11 Leaves entire or divided with the rachis and divisions commonly over 1 mm wide; native
12 Leaves mostly 1-3 times pinnately dissected, less pubescent on upper surface, very strongly aromatic; involucres glabrous or nearly so .... **A. michauxiana**
12 Leaves entire to sub-bipinnatifid, pubescence various, odor moderate; involucres tomentose to glabrate .... **A. ludoviciana**
6 Flowers at middle of head sterile, the ovary aborted
13 Leaves mostly entire, the lower rarely with 3-5 narrow segments .... **A. dracunculus**
13 Leaves mostly pinnatifid or dissected except the upper ones .... **A. campestris**

†**1. A. abrotanum** L. Sagebrush and disturbed areas, remaining from previous cultivation.

2. **A. arbuscula** Nutt.

    Leaves merely toothed or lobed less than halfway to base . . . . ssp. **arbuscula**

    Leaves mostly lobed about halfway or more to base . . . . ssp. **thermopola** Beetle

    Ssp. **arbuscula.** Frequent on west side of Snake River north of Moose.

    Ssp. **thermopola.** Frequent on volcanic soils especially near boundary of Yellowstone Park.

3. **A. biennis** Willd. Rare in disturbed areas.
4. **A. campestris** L. ssp. **borealis** (Pall.) Hall & Clements var. **scouleriana** (Bess.) Cronq. Rare in open places.
5. **A. cana** Pursh ssp. **viscidula** (Osterh.) Beetle. Frequent along the Snake River.
6. **A. dracunculus** L. Infrequent in disturbed areas.
7. **A. frigida** Willd. Infrequent in gravelly places.
8. **A. ludoviciana** Nutt.

    Main leaves entire or simply lobed . . . . . . var. **ludoviciana**

    Main leaves somewhat deeply parted or divided

    Involucre 3.5-4 mm long, 3-7 mm wide; leaves white-tomentose on both sides . . var **latiloba** Nutt.

    Involucre 3-3.5 mm long, 2.5-4 mm wide; leaves sometimes green on upper side . . . . var. **incompta** (Nutt.) Cronq.

    Var. **incompta.** Frequent in the mountains.

    Var. **latiloba.** Infrequent in open places.

    Var. **ludoviciana.** Rare in the lowlands.

9. **A. michauxiana** Bess. Rare in rocky canyons.
10. **A. norvegica** Fries ssp. **saxatilis** (Bess.) Hall & Clements. Frequent, alpine and subalpine.
11. **A. rothrockii** Gray. Rare on high mountain slopes. The relationship between *A. rothrockii* and *A. tridentata* ssp. *vaseyana* is still not clear. The treatment here is therefore tentative.
12. **A. scopulorum** Gray. Rare, alpine and subalpine.
13. **A. tridentata** Nutt. Very common in open places.
14. **A. tripartita** Rydb. Infrequent in open places.

ASTER Aster

Mostly perennial herbs. Lvs alt. Heads many flowered usually arranged in corymbs, panicles or racemes. Bracts strongly imbricated in several series. Achenes more or less flattened;

pappus of capillary bristles. Ray fls of varying shades of blue, purple or pink to white.

1 Plants with a taproot, no rhizomes or fibrous roots **A. alpigenus**
1 Plants with fibrous roots, often with creeping rhizomes
  2 Involucral bracts usually with a keeled midrib (use strong hand lens), usually purple tipped (at least the inner ones); rays 4-15
    3 Outer involucral bracts obtuse; ray flowers white or violet .... **A. glaucodes**
    3 Outer involucral bracts acute or acuminate
      4 Ray flowers deep violet; leaves 3-10 mm wide .. .. **A. perelegans**
      4 Ray flowers usually white, sometimes turning pink in age; leaves 1.5-3.5 cm wide .... **A. engelmannii**
  2 Involucral bracts without keeled midribs; rays usually more than 15
    5 Involucral bracts and peduncles with glandular hairs
      6 Leaves ovate to oval, sharply serrate to dentate .... **A. conspicuus**
      6 Leaves lanceolate, or linear, usually entire
        7 Leaves all sessile .... **A. campestris**
        7 Lower leaves with petioles .... **A. integrifolius**
    5 Involucral bracts and peduncles without glandular hairs
      8 Pubescence on stems and branchlets positioned in lines below the leaf bases and not uniform under the heads or confined to the inflorescence .... **A. hesperius**
      8 Pubescence evenly distributed or if in lines, then either uniform under the heads or sparse and limited to the inflorescence
        9 Outer involucral bracts much shorter than the inner, obtuse and not leaflike .... **A. chilensis**
        9 Outer involucral bracts nearly as long as the inner
          10 Middle stem leaves 1 cm wide or more, mostly less than 7 times as long as broad .... **A. foliaceus**
          10 Middle stem leaves mostly less than 1 cm wide and more than 7 times as long as wide .... **A. occidentalis**

1. **A. alpigenus** (T. & G.) Gray var. **haydenni** (Porter) Cronq. Frequent on the shores of alpine lakes and on some mountain summits.
2. **A. campestris** Nutt. Common along the Snake River.
3. **A. chilensis** Nees ssp. **adscendens** (Lindl.) Cronq. Common on the dry terraces above Pacific Creek.
4. **A. conspicuus** Lindl. Common on the north side of Signal Mt.
5. **A. engelmannii** (D. C. Eat.) Gray. Frequent around the shores of the valley lakes and in the major canyons up to 9,000 ft.
6. **A. foliaceus** Lindl. Common in sagebrush and open forest of the valley. A variable species difficult to separate into varieties.
7. **A. glaucodes** Blake. Frequent on talus in major canyons.
8. **A. hesperius** Gray var. **laetevirens** (Greene) Cronq. Common in the sedge grass meadows near the Oxbow Bend of Snake River.
9. **A. integrifolius** Nutt. Common throughout the valley floor and in some canyons.
10. **A. occidentalis** (Nutt.) (T. & G.) var. **occidentalis.** Common in the streamside forest.
11. **A. perelegans** Nels. & Macbr. Frequent in the outwash plain and extending up the mountains.

BALSAMORHIZA Balsamroot

Perennial herbs with thick tap-roots. Lvs mostly basal, large and long-petioled. Heads large and showy, mostly solitary. Ray fls yellow, pistillate. Receptacle chaffy. Achenes 4-angled; pappus none.

1 Leaves sagittate, margin entire .... **B. sagittata**
1 Leaves pinnately cleft or incised .... **B. macrophylla**

1. **B. macrophylla** Nutt. Large-leaf Balsamroot. Rare ½ mile north of Menor's Ferry.
2. **B. sagittata** (Pursh) Nutt. Arrowleaf Balsamroot. Abundant in sagebrush flats and on moraines of the valley.

BIDENS Beggar-tick

Annual herbs (in ours). Lvs mostly opp. Heads with yellow ray and disk fls. Involucral bracts in 2 series; receptacle flat and chaffy. Achenes 4-angled; pappus of a few bristles.

1. **B. cernua** L. Locally common on the margins of ponds and shallow lakes.

BRICKELLIA Brickellia

Perennial herbs (in ours). Lvs mostly alt, simple. Heads discoid; fls tubular and perfect. Involucral bracts imbricated in several series; receptacle flat, naked; corollas white or creamy. Achenes 10-ribbed.

1. **B. grandiflora** (Hook.) Nutt. Frequent on the lower portions of the canyon trails.

CARDUUS Thistle

Annual, biennial or perennial spiny herbs. Lvs alt, serrate to pinnately lobed, the stem generally winged by the decurrent leaf bases. Heads of disk fls only; involucral bracts imbricated in several series. Receptacle flat or convex, densely bristly. Corollas purple or reddish to white. Achenes glabrous. Similar to *Cirsium*, but differing chiefly in the merely minutely barbellate rather than plumose pappus bristles.

†**1. C. nutans** L. Nodding Thistle, Bristlethistle. Roadsides and disturbed sites; native of Europe.

CHAENACTIS Chaenactis

Biennial or perennial herbs (in ours). Lvs alt, entire to irregularly dissected. Heads of disk fls only, perfect. Involucral bracts narrow, equal or imbricated; receptacle naked; fls white or pink. Achenes club-shaped.

1 Plants alpine, low, perennials with 1 or usually several rosettes .... **C. alpina**
1 Plants seldom alpine, biennials or perennials, mostly erect (2 dm or more) .... **C. douglasii**

1. **C. alpina** (Gray) M. E. Jones. Frequent on the slopes of the higher peaks.
2. **C. douglasii** (Hook.) H. & A. Frequent on the Snake River.

CHRYSANTHEMUM Chrysanthemum

Annual or perennial herbs. Lvs alt. Both ray and disk fls

present; receptacle flat without chaff. Achenes ribbed; pappus absent. Ray fls white.

1 Heads solitary or few, the disk mostly 1-2 cm wide, the rays mostly 1-2 cm long .... **C. leucanthemum**
1 Heads several or numerous, the disk mostly 4-9 mm wide, the rays less than 1 cm long .... **C. parthenium**

†**1. C. leucanthemum** L. Oxeye Daisy. Infrequent on roadsides; native of Eurasia.
†**2. C. parthenium** (L.) Bernh. Feverfew. Disturbed ground; native of Europe.

CHRYSOPSIS Golden Aster

Annual or perennial herbs. Lvs alt, mostly entire. Ray and disk fls yellow. Involucral bracts in several series; receptacle flat, naked. Achenes flattened; pappus double, the inner of capillary bristles, the outer of short bristles.

1. C. villosa (Hook.) Nutt. Infrequent in the gravel of Snake River and Pilgrim Creek.

CHRYSOTHAMNUS Rabbitbrush

Shrubs. Lvs alt, narrow, entire. Heads with 5-30 yellow disk fls. Involucral bracts imbricated in 5 distinct vertical ranks; receptacle naked. Achenes terete or slightly angled; pappus of capillary bristles.

1 Branches glabrous or minutely puberulent, not at all tomentose .... **C. viscidiflorus**
1 Branches covered with a felt-like tomentum, scrape with fingernail to make sure .... **C. nauseosus**

1. C. nauseosus (Pall.) Britt. var. **petrophilus** Cronq. Common on the east side of the valley.
2. C. viscidiflorus (Hook.) Nutt. var. **stenophyllus** (Gray) Hall. Frequent, Blacktail Pond overlook and 2 miles east of Elk Reservoir.

CICHORIUM Chicory

Annual or perennial herbs. Lvs alt, entire to pinnatifid. Fls all ligulate and perfect, blue or white. Achene about 5-angled; pappus of scales. Milky juice.

†1. **C. intybus** L. Roadside weed; native of Eurasia.

CIRSIUM Thistle

Annual, biennial or perennial spiny herbs. Lvs alt, toothed to pinnatifid. Disk fls only, tubular and perfect; receptacle flat, bristly; corollas purple, red to yellow or white. Achenes quadrangular or flattened; pappus of plumose bristles.

A difficult genus that needs more study in Teton County.

1 Plants partly dioecious; head rather small, 1-1.5 cm high; introduced weed .... **C. arvense**
1 Plants with perfect flowers; head large, 2-5 cm high
  2 Leaves rough to the touch on the upper surface (scabrous); stem conspicuously winged by the decurrent leaf bases; introduced weed .... **C. vulgare**
  2 Leaves not at all scabrous, but with various types of pubescence; leaf bases not conspicuously decurrent; native to the area
    3 Plants well branched, prominently spined, not at all succulent; leaves more or less woolly below **C. subniveum**
    3 Plants differing in 1 or more ways from the above
      4 Plants spiny 1.5-8 dm tall, not succulent, seldom branched .... **C. tweedyi**
      4 Plants up to 12 dm tall; stems thick and succulent throughout, somewhat branched .... **C. scariosum**

†1. **C. arvense** (L.) Scop. Common in waste places.
2. **C. scariosum** Nutt. Common in wet meadows throughout the valley [*C. foliosum* (Hook.) DC.]
3. **C. subniveum** Rydb. Jackson's Hole Thistle. Fls pale pink-purple. Lower end of the valley.
4. **C. tweedyi** (Rydb.) Petr. Reported in the subalpine areas.
†5. **C. vulgare** (Savi) Tenore. Infrequent in disturbed sites.

CONYZA Horseweed

Similar to *Erigeron*, but the outer, ♀ fls of each head very numerous. Heads disciform or minutely radiate. Pappus of capillary bristles. Ray fls white or purplish.

†1. **C. canadensis** (L.) Cronq. Annual weed of disturbed ground.

CREPIS Hawk's Beard

Annual, biennial or perennial herbs with milky juice. Lvs alt or mostly basal, entire to pinnatifid. Heads arranged in cymes or panicules. Involucral bracts in 1 series or with a few short ones below. Achenes 10-20 ribbed; pappus of numerous capillary bristles. Ligulate fls yellow.

1 Plants annual; introduced weed .... **C. tectorum**
1 Plants perennial; not weedy
  2 Involucral bracts or lower part of stem or both, conspicuously bristly, but not at all glandular ... **C. modocensis**
  2 Involucral bracts and stem sparingly or not at all bristly
    3 Heads very narrow, mostly 5-10 flowered; 20-100 or more heads .... **C. acuminata**
    3 Heads broader, mostly 10-40 flowered; 3-40 heads .... **C. atrabarba**

1. C. acuminata Nutt. Infrequent in streambeds and under lodgepole.
2. C. atrabarba Heller. North slope of Signal Mt.
3. C. modocensis Greene. Frequent in sagebrush.
†**4. C. tectorum** L. Frequent along Snake River.

ERIGERON Daisy; Fleabane

Annual or perennial herbs. Lvs alt, entire, toothed or dissected. Head solitary, corymbose or paniculate. Involucral bracts narrow, equal, or somewhat imbricate; receptacle flat, naked. Ray fls blue, purple, pink or white, usually showy. Achenes 2-nerved; pappus of capillary bristles. This genus is closely related to *Aster* and not easy to separate into clearly defined taxa.

1 Leaves 1-3 times ternately dissected .... **E. compositus**
1 Leaves entire or slightly dentate or serrate
  2 Pistillate corollas very numerous, with very narrow, short, erect rays; involucral bracts glandular or hirsute or both
    3 Inner involucral bracts usually long attenuate; inflorescence more or less flat-topped; pappus reddish-brown .... **E. acris**

3 Inner involucral bracts only acute or acuminate; inflorescence like a raceme; pappus usually white . . . . **E. lonchophyllus**

2 Pistillate corollas few to numerous, the rays when present, well developed and spreading, not short, narrow and erect

4 Stem leaves usually lanceolate or ovate; plants when mature tall and erect, somewhat Aster-like

5 Ray flowers broad, 2-4 mm wide, occasionally less than 2 mm; subalpine meadows . . . . **E. peregrinus**

5 Ray flowers narrow, less than 1 mm wide

6 Stems with rather equal leaves, the upper leaves gradually reduced . . . . **E. speciosus**

6 Stems with unequal-sized leaves, the uppermost leaves strongly reduced . . . . **E. glabellus**

4 Stem leaves often much reduced, mostly linear or oblanceolate; plants scarcely Aster-like, mostly relatively low and often spreading

7 Hairs of stems appressed or ascending or rarely lacking

8 Involucral bracts glandular, not hairy; leaves glabrous . . . . **E. leiomerus**

8 Involucral bracts more or less hairy, sometimes glandular also; leaves hairy

9 Basal leaves usually with 3 veins; stems usually conspicuously reddish-purple at the base . . . . **E. eatonii**

9 Basal leaves with one definite vein; stems not conspicuously reddish-purple at the base . . . . **E. ochroleucus**

7 Hairs of the stem widely spreading

10 Disk corollas 2-3 mm long; rays 75-150 . . . . **E. divergens**

10 Disk corollas 3.5-5 mm long; rays 50-100 . . . . **E. pumilus**

1. **E. acris** L. var. **debilis** Gray. Rocky places in the mountains.
2. **E. compositus** Pursh. Rocky alpine meadows in South Cascade Canyon and Amphitheater Lake.
3. **E. divergens** T. & G. Common in the sagebrush flats and the Snake River bottom.

4. **E. eatonii** Gray. Frequent in rocks of major canyons.
5. **E. glabellus** Nutt. Frequent along the Snake River.
6. **E. leiomerus** Gray. Infrequent in the subalpine and alpine meadows above 9,000 ft.
7. **E. lonchophyllus** Hook. Infrequent along streams north of Jackson Lake Dam.
8. **E. ochroleucus** Nutt. Frequent in the sagebrush areas near the Snake River.
9. **E. peregrinus** (Pursh) Greene ssp. **callianthemus** (Greene) Cronq. Frequent on the shores of sub-alpine lakes above 9,000 ft.
10. **E. pumilus** Nutt. ssp. **intermedius** Cronq. Common in the sagebrush flats throughout the valley.
11. **E. speciosus** (Lindl.) DC. var. **macranthus** (Nutt.) Cronq. Common in the sagebrush flats and along the Snake River.

ERIOPHYLLUM Eriophyllum

Perennial tomentose herbs (in ours). Lvs mostly alt, entire to variously cut. Usually several stems from the base; heads long-pedunculate; ray fls yellow. Achenes glabrous to glandular or hairy; pappus of scales or lacking.

1. **E. lanatum** (Pursh) Forbes var. **integrifolium** (Hook.) Smiley. Common in sagebrush flats and on the moraines of the valley.

GNAPHALIUM Cudweed; Everlasting

Annual to perennial white-woolly herbs. Lvs alt, entire. Heads with disk fls only, yellow or whitish. Involucral bracts imbricated in several series; receptacle naked. Achene terete; pappus of capillary bristles. A critical genus with many of the spp. being separated on trivial differences.

1 Plants annual; usually less than 3 dm tall, usually much branched . . . . **G. palustre**
1 Plants from a short lived perennial taproot, mostly 2-9 dm tall, only moderately branched . . . . **G. microcephalum**

1. **G. microcephalum** Nutt. Rare in Snowshoe Canyon.
2. **G. palustre** Nutt. Common along Snake River.

GRINDELIA Gumweed

Annual, biennial or perennial herbs, sometimes woody at the base. Lvs alt. Ray and disk fls yellow. Involucral bracts resinous or gummy, imbricated; receptacle flat or convex. Achenes 4-5 ribbed; pappus of 2-several awns.

†1. **G. squarrosa** (Pursh) Dunal. Common along roads and disturbed ground.

GUTIERREZIA Snakeweed

Herbs or low shrubs. Lvs alt, small and linear. Ray and disk fls yellow. Involucral bracts strongly imbricated; receptacle small, naked; pappus of several scales or awns, or absent. Achenes terete, usually with several nerves.

1. **G. sarothrae** (Pursh) Britt. & Rusby. Dry, open sites on the National Elk Refuge.

HAPLOPAPPUS Goldenweed

Annual or perennial herbs or shrubs. Lvs usually alt or occasionally basal. Ray and disk fls usually yellow. Involucral bracts strongly imbricated; receptacle naked and flat. Achene angled to terete; pappus of capillary bristles.

1 Plants shrubby, low, branching, not at all caespitose .... **H. suffruticosus**
1 Plants herbaceous, sometimes with a woody caudex, often densely caespitose
  2 Plants cushion-forming with much branched caudex, rarely over 2 dm tall .... **H. acaulis**
  2 Plants not cushion-forming, caudex simple or moderately branched, often well over 2 dm tall ...... **H. uniflorus**

1. **H. acaulis** (Nutt.) Gray. Infrequent west side of Blacktail Butte.
2. **H. suffruticosus** (Nutt.) Gray. Infrequent in major canyons and Alaska Basin.
3. **H. uniflorus** (Hook.) T. & G. Dry, open places, east side of park.

HELENIUM Sneezeweed

Annual or perennial herbs. Lvs alternate, simple. Involucral bracts in 2-3 series; heads generally radiate; ray fls yellow or yellow orange (in ours), cuneate, 3-lobed; disk fls numerous; anthers slightly sagittate; pappus of several series; receptacle convex or conic, naked.

1. **H. hoopesii** Gray. Locally frequent, Teton Natl. Forest, 4 miles east of Blackrock Ranger Station.

HELIANTHELLA Mountain Sunflower

Perennial herbs, 3-10 dm tall. Lvs mostly opp, simple and entire. Ray and disk fls (in ours) yellow. Involucral bracts green, in 2 series; receptacle chaffy. Achenes flattened; pappus of several short scales and commonly 2 slender awns.

1 Leaves with 2 prominent pairs of lateral veins; involucral bracts ovate or lanceolate; receptacular bracts soft and scarious .... **H. quinquenervis**
1 Leaves more or less triple veined; involucral bracts mostly lance-linear; receptacular bracts firm .... **H. uniflora**

1. **H. quinquenervis** (Hook.) Gray. Morainal soils throughout west side of valley.
2. **H. uniflora** (Nutt.) T. & G. Frequent on morainal soils of the valley and extending into the canyons.

HELIANTHUS Sunflower

Annual or perennial herbs. Lvs mostly opp, simple. Heads solitary or corymbose. Involucral bracts imbricated in several series; receptacle flat to convex, chaffy throughout. Achenes moderately compressed; pappus of 2 readily deciduous awns.

†1. **H. annuus** L. Frequent in waste places.

HIERACIUM Hawkweed

Perennial herbs with milky juice. Lvs alt or all basal, entire or more or less toothed. Fls all ligulate and perfect, mostly yellow, sometimes red-orange, rarely white. Receptacle naked. Achenes terete or prismatic; pappus of capillary bristles. A critical genus in which the species are poorly defined because of apomixis and polyploidy.

1 Flowers white; stellate hairs absent .... **H. albiflorum**
1 Flowers yellow; stellate hairs always present
  2 Leaves glabrous or sometimes short-hairy; plants of alpine areas .... **H. gracile**
  2 Leaves generally more or less long-hairy
    3 Herbage moderately setose below, subglabrous and often glaucous above .... **H. scouleri**
    3 Herbage definitely long setose above as well as below, not at all glaucous .... **H. cynoglossoides**

1. **H. albiflorum** Hook. Frequent in shade of lodgepole pine forest.
2. **H. cynoglossoides** Arv. Infrequent in lodgepole pine forest.
3. **H. gracile** Hook. Frequent in the subalpine and alpine meadows.
4. **H. scouleri** Hook. Frequent in the sagebrush community.

HYMENOXYS Hymenoxys

Perennial or annual aromatic herbs. Lvs alt or basal, entire to pinnatifid. Ray and disk fls yellow; receptacle conical, naked. Achenes mostly 5-angled; pappus of 5-7 awned scales.

1. **H. grandiflora** (T. & G.) Parker. Frequent on talus slopes above 9,000 ft.

IVA Poverty-weed

Annual or perennial herbs, lvs opp (or upper lvs alt). Heads spicate, racemose or solitary in the axils. Involucral bracts imbricated; receptacle chaffy. Achenes flattened; pappus none. Fls greenish-white.

†**1. I. xanthifolia** Nutt. Sumpweed. Common in waste places.

LACTUCA Lettuce

Annual, biennial or perennial, leafy-stemmed herbs with milky juice. Lvs alt, entire to pinnatifid. Fls all ligulate, perfect, yellow, blue or white. Receptacle naked. Achenes compressed, beaked, surmounted by a pappus of numerous capillary bristles.

1 Plants perennial; head relatively large and showy; beak less than ½ as long as the achene; flowers blue or violet .... **L. pulchella**

1 Plants annual or biennial; heads relatively small; beak ½ as long as the achene or longer; flowers yellow, but often turning blue in age . . . . **L. serriola**

1. L. pulchella (Pursh) DC. Blue Lettuce. Infrequent along main highway south of Timbered Island.

†**1. L. serriola** L. Prickly Lettuce. Frequent in streambeds and disturbed sites.

MACHAERANTHERA

Annual to perennial herbs with a taproot system. Lvs alt. Heads solitary to many, small to large. Bracts in several series, papery or coriaceous toward base, greenish toward tip. Achenes more or less compressed, mostly several nerved; pappus of numerous, unequal, barbellate bristles. Ray fls bluish to lavender or whitish.

1. M. canescens (Pursh) Gray. Common in the sagebrush flats of the valley [*Aster canescens* Pursh].

MADIA Tarweed

Annual (in ours) herbs with glandular hairs and an oily odor. Lvs usually alt, at least the lower lvs opposite, entire or slightly toothed. Heads radiate, yellow. Involucral bracts in a single series and usually enclosing the ray achene; pappus usually none.

†**1. M. glomerata** Hook. Locally common in disturbed sites of the valley.

MATRICARIA Mayweed

Annual or perennial herbs. Lvs alt, pinnatifid or pinnately dissected. Ray or disk fls, the rays white. Receptacle naked, conic or elongate; disk fls yellow. Achenes ribbed; pappus usually absent.

1 Heads with white ray flowers . . . . **M. chamomilla**
1 Heads with disk flowers only . . . . **M. matricarioides**

†**1. M. chamomilla** L. Rare on roadside cuts.

†**2. M. matricarioides** (Less.) Porter. Pineapple Weed. Common in waste places throughout the valley.

MICROSERIS Microseris

Annual or perennial, taprooted herbs with milky juice. Lvs alt (or all basal), entire to pinnatifid. Heads 1-many on the ends of naked peduncles. Fls all ligulate and perfect, yellow. Achenes 8-10 ribbed; pappus of slender white scales, each terminating in a plumose bristle.

1. **M. nutans** (Geyer) Schultz-Bip. Infrequent in Lupine Meadows south of Jenny Lake.

RATIBIDA Coneflower

Biennial or perennial herbs. Lvs alt, pinnatifid. Ray fls usually 3-7, yellow or sometimes purple; disk fls numerous, brown; receptacle columnar. Achenes compressed; pappus of 1 or 2 teeth.

†1. **R. columnifera** (Nutt.) Woot. & Standl. Rare on roadside cuts of the valley.

RUDBECKIA Coneflower

Annual or perennial herbs. Lvs alt, entire to pinnatifid. Ray fls lacking (in ours); disk fls purplish brown. Receptacle enlarged, conic or columnar, chaffy throughout. Achenes quadrangular; pappus a short, often toothed or irregular crown, or none.

1. **R. occidentalis** Nutt. Rayless Coneflower. Locally frequent around streams and lakes.

SENECIO Groundsel

Annual, biennial or perennial herbs (in ours). Lvs alt or all basal, entire to divided. Heads radiate or sometimes discoid, yellow to orange. Involucral bracts in 1 series; receptacle naked, flat or convex. Achenes terete; pappus of numerous capillary bristles. One of the largest genera, and difficult to separate into distinct taxa.

1 Stems more or less leafy throughout; leaves about the same shape
  2 Stem leaves, at least most of them, sessile with broad, more or less clasping bases .... **S. crassulus**
  2 Stem leaves, except some of the upper ones, petiolate, or tapering to narrow petiole-like bases

3 Lower leaves triangular with deltoid to cordate base .... **S. triangularis**

3 Leaves not triangular, but tapering to the base .. **S. serra**

1 Stems with leaves becoming definitely smaller and usually different shape higher on the stem; basal leaves often tufted

4 Flowering heads 1-2, nodding; plants above timberline .... **S. amplectens**

4 Flowering heads 1-many, usually erect; plants of various habitats

5 Plants glabrous from seedling stage on except for a little pubescence at the base and in the leaf axils

6 Leaves entire or denticulate, not at all pinnatifid, lobed, wavy or crenate .... **S. hydrophilus**

6 Leaves, or some of them, pinnatifid, crenate or wavy

7 Leaves relatively thin and lax; plants often at lower elevations .... **S. pauperculus**

7 Leaves relatively thick and firm; plants at mid to high elevations

8 Stem leaves well developed, middle and lower leaves clasping .... **S. dimorphophyllus**

8 Stem leaves progressively reduced, not clasping .... **S. streptanthifolius**

5 Plants more or less pubescent at flowering time

9 Plants conspicuously pubescent at flowering time

10 Stems tall (4 dm or more); moderate elevations in the mountains .... **S. sphaerocephalus**

10 Stems mostly 1-3 dm tall; timberline or above (pubescence sometimes partly deciduous) ... . **S. canus**

9 Plants sparingly pubescent at flowering time

11 Pubescence of coarse multicellular hairs (use strong hand lens) .... **S. integerrimus**

11 Pubescence of fine floccose hairs; septa not apparent .... **S. sphaerocephalus**

1. **S. amplectens** Gray. Frequent on limestone, Moose Basin and So. Teton Canyon.
2. **S. canus** Hook. Infrequent in rocks of tundra above 10,000 ft.
3. **S. crassulus** Gray. Frequent in quaking aspen in area of Two Ocean Lake, also major canyons above 9,400 ft.

4. **S. dimorphophyllus** Greene var. **paysonii** T .M. Barkley. Frequent in main couloir on the east face of Symmetry Spire, 9,000 ft.
5. **S. hydrophilus** Nutt. Locally frequent in the ponds east of Colter Bay.
6. **S. integerrimus** Nutt. Common in the sagebrush flats throughout the valley.
7. **S. pauperculus** Michx. Frequent on the west side of the valley.
8. **S. serra** Hook. Frequent on the dry bottoms of Cottonwood Creek.
9. **S. sphaerocephalus** Greene. Disturbed sites in Willow Flats north of Jackson Lake Dam.
10. **S. streptanthifolius** Greene. (*S. cymbalarioides* Nutt.) Oxbow Bend area.
11. **S. triangularis** Hook. Common along streams of the major canyons.

SOLIDAGO Goldenrod

Fibrous-rooted perennial herbs with a rhizome or a caudex. Lvs alt, entire or variously toothed. Heads radiate (in ours), fls yellow. Involucral bracts imbricated in several series; receptacle flat and naked. Achenes several-nerved; pappus of capillary bristles. A critical genus in which the species are difficult to define.

1 Plants with a well developed creeping rhizome; leaves are about equal in length throughout the stem
  2 Stem glabrous below the flowers; leaf blades mostly glabrous .... **S. missouriensis**
  2 Stems with a fine pubescence, at least above the middle; leaves with some pubescence
    3 Plants 1-6 dm; ray flowers about 8 mm long .... **S. mollis**
    3 Plants 5-20 dm; ray flowers about 13 mm long .... **S. canadensis**
1 Plants with a short and stout rhizome; lower leaves larger than those on the stem above
  4 Leaves densely covered with short spreading hairs .. **S. nana**
  4 Leaves glabrous except for the occasional ciliate margins
    5 Ray flowers mostly about 13; petioles of lower leaves with a ciliate margin .... **S. multiradiata**

5 Ray flowers mostly about 8; petioles lacking a ciliate margin .... **S. spathulata**

1. **S. canadensis** L. var. **salebrosa** (Piper) M. E. Jones. Frequent along the Snake River in disturbed areas.
2. **S. missouriensis** Nutt. Frequent in the streamside forest along the Snake River.
3. **S. mollis** Bartl. Infrequent near Jenny Lake.
4. **S. multiradiata** Ait. Frequent along the Snake River and lower portions of the major canyons.
5. **S. nana** Nutt. Dry hillside of Cascade Canyon.
6. **S. spathulata** DC. Common in the meadows east and north of Jackson Lake Dam.

STEPHANOMERIA Wirelettuce; Skeletonweed

Annual or perennial herbs with milky juice. Lvs alt, mostly small and often scalelike. Flowers all ligulate and perfect, pink or occasionally white. Achenes 5-ribbed; pappus of slender bristles, plumose above.

1. **S. tenuifolia (Torr.)** Hall. Frequent on wet sand and rocky bars of the Snake River.

TANACETUM Tansy

Annual or perennial herbs, often aromatic. Lvs alt, 1-3 pinnatifid. Heads (in ours) discoid, numerous, from 20-200; corollas yellow. Achenes mostly 5-ribbed; pappus a short crown or none.

†1. **T. vulgare** L. Frequent along Snake River below Jackson Lake Dam, remaining from previous cultivation.

TARAXACUM Dandelion

Taprooted perennial herbs, stems scapose, milky juice. Lvs basal, entire to pinnatifid. Flowers all ligulate and perfect, yellow. Achenes with a prominent beak; pappus of capillary bristles. Hybridization, polyploidy and apomixis have combined to make this a critical genus of confusing species.

1 Achenes becoming red to reddish brown or reddish purple at maturity; leaves tending to be deeply cut for their whole length .... **T. laevigatum**

1 Achenes straw-colored to brown; leaves usually less deeply cut .... **T. officinale**

†**1. T. laevigatum** (Willd.) DC. Frequent on disturbed ground throughout the valley.
†**2. T. officinale** Weber. Abundant throughout the valley.

TETRADYMIA Horsebrush

Low branching shrubs. Lvs alt, narrow and entire. Stems more or less canescent. Heads discoid, yellow, 4-9 flowered; receptacle naked. Achenes terete, 5-nerved; pappus of white capillary bristles.

1. T. canescens DC. Locally frequent in the sagebrush of the valley.

TOWNSENDIA Daisy; Townsendia

Annual, biennial or perennial herbs with deep tap roots. Lvs alt or all basal, entire. Heads solitary or few, many flowered, involucral bracts, narrow, partially imbricate; ray flowers blue, purple to pinkish or white; disk flowers numerous, perfect; achenes flattened, usually pubescent; pappus of a single series of bristle-like scales.

1 Ray flowers white or pinkish; achenes persistently hairy at the base; plant of the valley .... **T. exscapa**
1 Ray flowers blue or violet to occasionally white; achenes glabrous to subglabrous; montane plant .... **T. montana**

1. T. exscapa (Rich.) Porter. Infrequent in dry, barren sites on on the east side of the valley.
2. T. montana Jones, var. **montana.** Frequent, rocky soil, north ridge of Teton Canyon, also Teton Pass.

TRAGOPOGON Goatsbeard

Annual, biennial or perennial taprooted herbs with milky juice. Lvs alt, linear, entire and clasping. Heads solitary at the ends of the branches. Flowers all ligulate and perfect, yellow or purple. Achenes linear, 5-10-nerved, usually beaked; pappus of a single series of plumose bristles.

1 Heads yellow .... **T. dubius**
1 Heads purple .... **T. porrifolius**

†1. **T. dubius** Scop. Frequent in disturbed sites throughout the valley.
†2. **T. porrifolius** L. Frequent in disturbed sites.

VIGUIERA Goldeneye

Perennial herbs. Lvs opp below, alt above; stems several. Heads radiate, yellow; ray flowers neutral; receptacle chaffy throughout, bracts clasping the achenes. Pappus of 2 awns and several scales or none.

1. **V. multiflora** (Nutt.) Blake var. **multiflora.** Common along main highway north of Jackson Lake Lodge.

WYETHIA Mulesears

Taprooted, leafy-stemmed, perennial herbs. Lvs alt, simple. Heads large, solitary or several, radiate. Ray flowers yellow or white; disk flowers light yellow; receptacle broadly convex, chaffy throughout, the bracts clasping the achenes. Pappus of unequal scales.

1 Ray flowers yellow .... **W. amplexicaulis**
1 Ray flowers white or pale creamy .... **W. helianthoides**

1. **W. amplexicaulis** Nutt. Locally common west of Cattleman's Bridge.
2. **W. helianthoides** Nutt. Locally common in moist meadows, Highway 89, 4 miles north of Colter Bay.

BERBERIDACEAE Barberry Family

Shrubs (in ours). Lvs alt, simple/divided, usually no stip. Infl cymose/racemose. Fls bisex, reg, ov superior. Ca 4-6, Co 4-6, S 4-18, P 1. Capsule/berry.

BERBERIS Barberry; Oregongrape; Mahonia

Evergreen shrubs with yellow inner bark. Lvs alt, leathery. Fls racemose; perianth has several whorls of parts, yellow; anthers with 2 uplifting trap doors for pollen release; berry 1-several-seeded.

1. **B. repens** Lindl. [*Mahonia repens* G. Don.]. Frequent around the shores of the valley lakes. Fruits are edible.

BETULACEAE Birch Family

Woody. Lvs alt, simple, stip. Infl catkin. Fls unisex, ♂ Per 4*, S 2-4; ♀ Per 0, P 2*, ov inferior. Nut, nutlet or small samara.

1 Female catkins soft or not cone-like, scales deciduous with fruits; leaves singly serrate .... **Betula**
1 Female catkins hardened and cone-like, scales persisting after shedding of fruits; leaves doubly serrate .... **Alnus**

ALNUS Alder

Deciduous trees or shrubs, plants monoecious. Lvs alt, simple, dentate to lobed. Fls unisexual in catkins, female catkin hard and cone-like, each with 2 naked fls.

1. A. incana (L.) Moench. [*A. tenuifolia* Nutt.] Frequent in moist places around the shores of the valley lakes and along the Snake River.

BETULA Birch

Deciduous trees or shrubs, plants monoecious. Lvs alt, usually serrate margins. Both staminate and pistillate fls in catkins; male catkin pendulous, 3 fls per cluster, each with 2-3 perianth segments, usually 2 stamens. Female catkins usually erect, 3-lobed bracts subtending 2-3 naked fls; fruit a winged samara.

1 Shrubs up to 4 m tall; young twigs with dense short hairs and glandular warts .... **B. glandulosa**
1 Shrubs 5-18 m tall or trees; young twigs glabrous, often not warty .... **B. occidentalis**

1. B. glandulosa Michx. Infrequent in meadows and wet places of the valley and some canyons.
2. B. occidentalis Hook. Rare on the south end of Blacktail Butte.

BORAGINACEAE Borage Family

Herbs. Lvs alt, simple, no stip. Infl often helicoid cymes. Fls bisex, reg, ov superior. Ca 5*, Co 5*, S 5, P 2*, ov deeply 4-lobed. Nutlets.

1 Flowers reddish-purple .... **Cynoglossum**
1 Flowers colored other than above

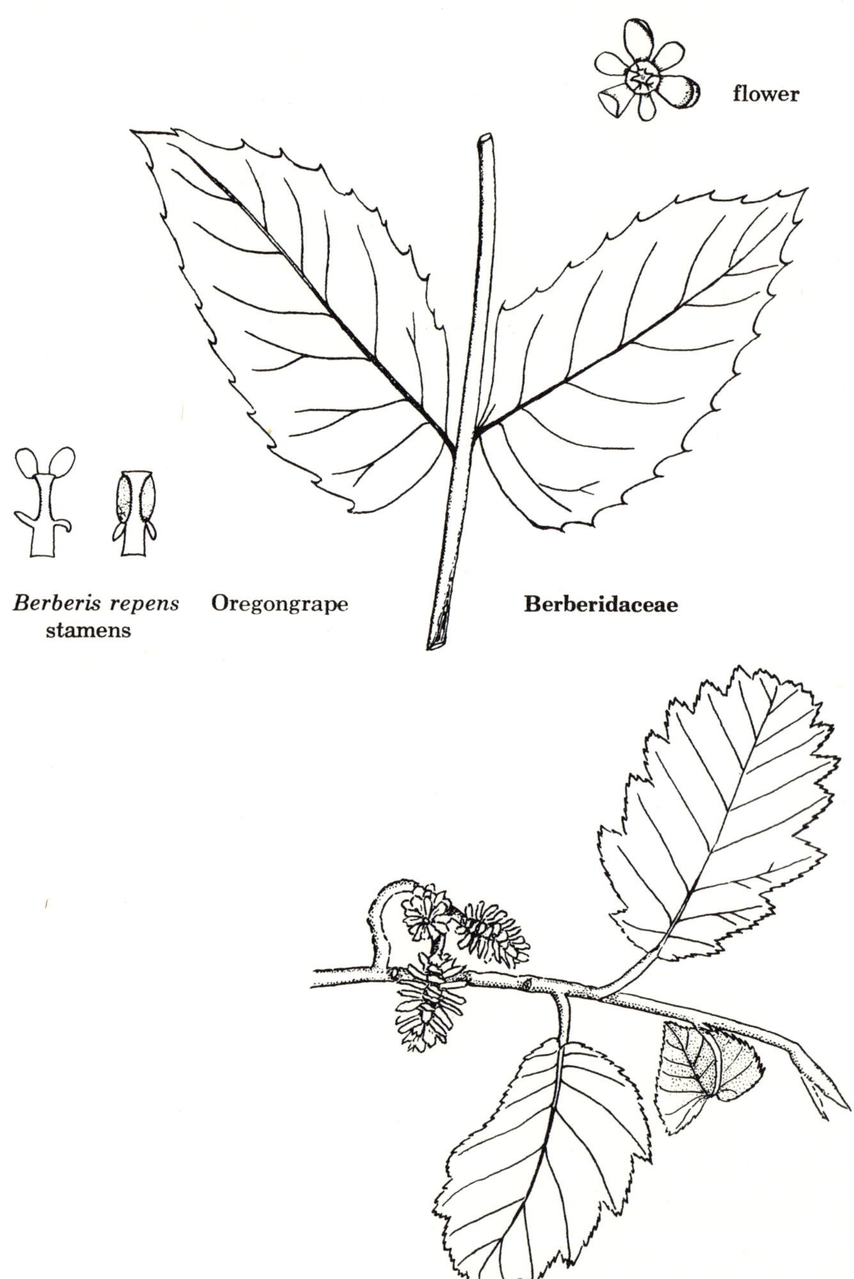

Alnus incana Alder

Betulaceae

2 Flowers yellow or orange
 3 Plants weedy, annuals with stiff hirsute pubescence; flowers lacking bracts .... **Amsinckia**
 3 Plants naturally occurring, perennials with appressed pubescence; flowers with bracts .... **Lithospermum**
2 Flowers colored other than yellow or orange
 4 Plants cushion-like; alpine perennials; flowers blue .... **Eritrichium**
 4 Plants erect and leafy-stemmed; various habitats; flower color various
  5 Plants weak and fragile; lower stem leaves opposite; wet habitats .... **Plagiobothrys**
  5 Plants robust; stem leaves alternate; various habitats
   6 Nutlets with glochidiate prickles
    7 Plants annual; flowers inconspicuous; in fruit the style surpassing the nutlets ... **Lappula**
    7 Plants perennial; flowers conspicuous; in fruit the nutlets surpassing the style .. **Hackelia**
   6 Nutlets without glochidiate prickles
    8 Flowers white .... **Cryptantha**
    8 Flowers blue or pink
     9 Corolla tubular; nutlets are generally wrinkled .... **Mertensia**
     9 Corolla salverform or broadly funnelform; nutlets smooth and shining .. **Myosotis**

AMSINCKIA Fiddle-neck; Tarweed

Bristly-hairy annuals from a taproot. Lvs alt, small and narrow. Flowers borne in helicoid, false racemes or spikes; corolla yellow or orange, the fornices lacking; nutlets more or less triangular.

†1. **A. menziesii** (Lehm.) Nels. & Macbr. Frequent on roadsides and abandoned fields on road to Gros Ventre Slide.

CRYPTANTHA Cryptantha

Annual or perennial herbs. Lvs alt, linear to spatulate. Flowers borne in helicoid false spikes; calyx cleft to the base; corolla white, fornices often yellow; nutlets smooth or rough with a narrow enlongate scar. A difficult genus with vague boundaries between taxa.

1 Nutlets obliquely compressed, with an off center scar near one margin . . . . **C. affinis**
1 Nutlets symmetrical, the scar median on the ventral face
  2 Fruiting calyx 2-4.5 mm long . . . . **C. watsonii**
  2 Fruiting calyx 4-8 mm long . . . . **C. torreyana**

1. C. affinis (Gray) Greene. Common in waste places throughout the valley.
2. C. torreyana (Gray) Greene. Frequent, dry, open sites of the valley.
3. C. watsonii (Gray) Greene. Frequent in disturbed sites at Colter Bay Village.

CYNOGLOSSUM Hound's tongue

Annual or perennial herbs from taproot. Lvs alt, entire. Fls borne in false racemes or mixed panicles; calyx deeply cleft; corolla reddish purple, funnelform or salverform, fornices present; nutlets covered with short glochidiate prickles.

†**1. C. officinale** L. Infrequent along roadsides of the valley.

ERITRICHIUM Alpine Forget-me-not

Dwarf, cushion-like perennial herbs. Lvs densely crowded on numerous short shoots, more or less hairy. Flowers borne in cymose clusters; corolla blue usually with a yellow center, salverform, fornices present; anthers included in tube; nutlets 1-4.

1. E. nanum (Vill.) Schrad. var. **elongatum** (Rybd.) Cronq. Locally frequent in the mountains above 10,000 ft.

HACKELIA Stickseed; Wild Forget-me-not

Biennial or perennial, taprooted herbs. Lvs alt, linear or oblong. Flowers borne in false racemes which tend to elongate in age; calyx deeply cleft; corolla white or blue, fornices well developed; stamens included; nutlets equipped with glochidiate prickles.

1 Corolla white, sometimes streaked with pale blue
  2 Corolla marked with light blue; pubescence on the stem mostly appressed . . . . **H. patens**

2 Corolla not streaked with blue; pubescence on the stem appressed or often spreading .... **H. diffusa**

1 Corolla blue

3 Plants biennial or perennial, usually with a single stem; intramarginal prickles lacking .... **H. floribunda**

3 Plants perennial, mostly with several stems; intramarginal prickles present .... **H. micrantha**

1. **H. diffusa** (Lehm.) I. M. Johnst. Frequent in aspen groves along the Snake River.
2. **H. floribunda** (Lehm.) I. M. Johnst. Frequent in meadows, stream banks, and other moist places, up to 9,000 ft.
3. **H. micrantha** (Eastw.) J. L. Gentry. Frequent along the Snake River.
4. **H. patens** (Nutt.) I. M. Johnst. Common in dry open sites, often with big sagebrush.

LAPPULA Stickseed

Annual, taprooted herbs. Lvs alt, linear. Fls borne in bracteate false racemes; calyx cleft to the base; corolla blue or white, mostly inconspicuous, well developed fornices; stamens included; nutlets with glochidiate prickles.

1. **L. redowskii** (Hornem.) Greene. Disturbed sites along the Snake River.

LITHOSPERMUM Puccoon; Stoneseed

Annual or perennial herbs. Lvs alt, linear. Fls borne in leafy cymes; calyx deeply cleft; corolla yellow to white, funnelform to salverform; stamens included to partly exserted; nutlets smooth to wrinkled.

1. **L. ruderale** Dougl. Frequent along the Snake River and extending out into the sagebrush flats.

MERTENSIA Bluebells

Perennial herbs from taproots. Lvs alt, entire and somewhat succulent. Flowers in bractless cymes; calyx cut at least to the middle; corolla blue, occasionally pink or white, often funnelform; fornices present; stamens often included; nutlets mostly wrinkled.

1 Plants usually less than 4 dm tall; blooming early in the spring; cauline leaves usually without conspicuous lateral veins .... **M. oblongifolia**

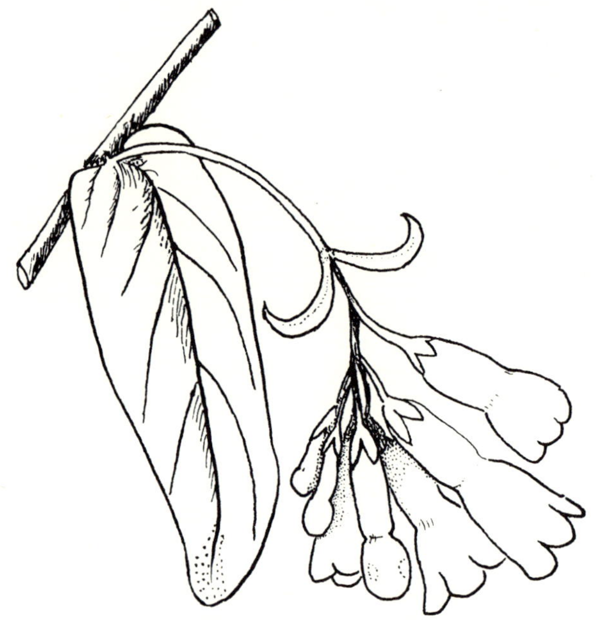

Mertensia ciliata Bluebells

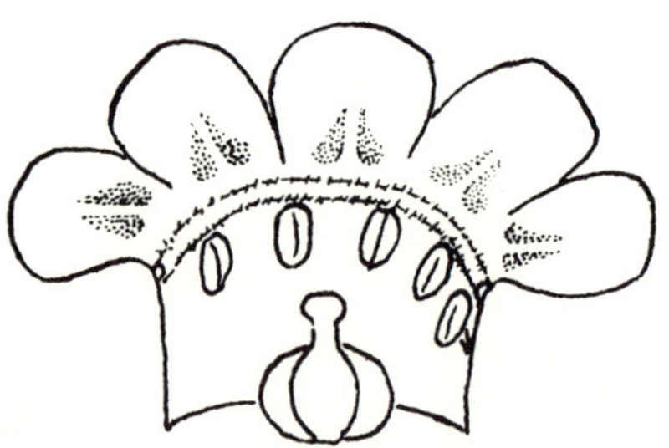

Hackelia patens Stickseed

Boraginaceae

1 Plants usually more than 4 dm tall, blooming in late spring and summer; cauline leaves with conspicuous lateral veins .... **M. ciliata**

1. **M. ciliata** (James) G. Don. Common along stream banks and wet meadows up to 9,000 ft.
2. **M. oblongifolia** (Nutt.) G. Don. Common in the sagebrush flats throughout the valley.

MYOSOTIS Forget-me-not

Annual to perennial herbs. Lvs alt, entire, linear. Fls borne in helicoid false racemes; calyx with a distinct tube; corolla mostly blue, fornices evident; nutlets smooth.

1 Calyx with appressed stiff hairs; corolla limb 2-5 mm wide .... **M. scorpioides**
1 Calyx with spreading uncinate hairs; corolla limb 4-8 mm wide .... **M. sylvatica**

1. **M. scorpioides** L. Locally frequent, cobblestone river bottom south of Park Hdqts.
2. **M. sylvatica** Hoffm. var. **alpestris** (F. W. Schmidt) Koch. Infrequent at moderate to high elevations in the mountains.

PLAGIOBOTHRYS Plagiobothrys

Annual or perennial herbs. Lvs mostly alt, but the lower often opp, linear to oblong. Fls borne in helicoid false racemes; calyx cleft to below the middle; corolla white, salverform, fornices evident; stamens included; nutlets smooth or rough.

1. **P. scouleri** (H. & A.) Johnst. var. **penicillatus** (Greene) Cronq. Popcorn Flower. Locally frequent in pond east of Colter Bay Fire Cache.

BRASSICACEAE (CRUCIFERAE) Mustard Family

Herbs. Lvs usually alt, simple/divided, no stip. Infl racemose/corymbose. Fls bisex, reg, ov superior. Ca 4, Co 4, S 4 + 2, P 2*. Silicle/silique.

Key based on vegetative, flower and mature fruit characters

1 Stems with clasping leaves
  2 Leaves simple and entire .... **Thlaspi**

2 Leaves variously toothed or dissected
3 Pubescence of branched or stellate hairs
4 Petals white
5 Fruit many times as long as wide .... **Arabis**
5 Fruit more or less obcordate .... **Capsella**
4 Petals pale yellow .... **Camelina***
3 Pubescence of simple hairs or none
6 Fruits orbicular or obovate, often winged at the top; seed one per locule .... **Lepidium**
6 Fruits not as above; several seeds per locule
7 Fruit oval to linear, not more than 10 mm long; seeds conspicuously notched near micropyle .... **Rorippa**
7 Fruit elongated, more than 10 mm long, seeds not notched
8 Margin of upper cauline leaves entire .. **Brassica**
8 Margin of upper cauline leaves conspicuously toothed .... **Barbarea**
1 Stems without clasping leaves
9 Petals pink, purple to magenta; silique not dehiscent .... **Chorispora**
9 Petals not as above; fruits dehiscent
10 Petals white
11 Plants aquatic; rooting at the nodes ..... **Nasturtium**
11 Plants terrestial; not rooting at the nodes
12 Leaves entire to dentate .... **Draba**
12 Leaves dissected or compound
13 Plants cushioned perennials with branching caudex .... **Smelowskia**
13 Plants annual to perennial with rhizomes .... **Cardamine**
10 Petals mostly yellow, sometimes cream to orange or reddish
14 Fruits siliques
15 Leaves simple, entire to dentate, style rather prominent .... **Erysimum**
15 Leaves 1-3 pinnately compound; style inconspicuous .... **Descurainia**
14 Fruits silicles

*Genus not yet represented in the flora, but the weedy species *C. microcarpa* Andrz. is expected soon.

16 Silicles flattened, style very short .... **Alyssum**
16 Silicles not flattened, usually appearing inflated; style conspicuous and persistent
  17 Silicles with halves separate and appearing as twins .... **Physaria**
  17 Silicles not twin-like .... **Lesquerella**

ALYSSUM Alyssum

Annual or perennial herbs. Lvs simple, entire to serrate. Fls borne in elongate, bractless racemes; sepals persistent to quickly deciduous; petals yellow or cream; fruit an ovate to elliptic silicle, strongly compressed. Plants with stellate pubescence.

1 Fruit with stellate pubescence; sepals persisting until the fruit is nearly mature .... **A. alyssoides**
1 Fruit glabrous; sepals deciduous soon after anthesis ... . **A. desertorum**

†1. **A. alyssoides** L. Locally frequent in disturbed sites.
†2. **A. desertorum** Stapf. An introduced weed from Europe spreading in our area.

ARABIS Rock Cress

Annual, biennial or perennial herbs. Lvs basal or alt, often clasping, entire or dentate. Fls in simple to branched racemes; sepals erect, greenish to purple; petals white to pink, red or purple; siliques linear, readily dehiscent. This is a difficult genus and immature specimens are often impossible to identify.

1 Siliques erect, not over 2 mm broad; flowers usually white but sometimes slightly pinkish to lavender; seeds usually wingless
  2 Cauline leaves not auriculate (ear-shaped appendages); plants usually less than 3-4 dm tall .... **A. nuttallii**
  2 Cauline leaves usually auriculate; plants mostly well over 3 dm tall, strongly hirsute at the base
    3 Seeds in one row in each chamber .... **A. hirsuta**
    3 Seeds in 2 rows in each chamber .... **A. glabra**
1 Siliques erect to reflexed, frequently over 2 mm broad; flowers white to pink or purple; seeds usually winged, the wing often over 0.3 mm wide

4 Plants mostly under 3 dm tall; seeds in one row in each chamber; stems usually glabrous; subalpine to alpine .... **A. lyallii**

4 Plants mostly well over 3 dm tall and in other ways not as above

5 Leaves usually densely pubescent and often grayish with small branching hairs (dendritic) on lower leaves; petals 4-6 mm long .... **A. cobrensis**

5 Leaves greenish to grayish, usually not densely pubescent; petals 5-12 mm long

6 Siliques strictly erect; seeds in 2 rows in each chamber .... **A. drummondii**

6 Siliques pendulous; seeds in one row in each chamber .... A. **holboellii**

1. **A. cobrensis** M. E. Jones. Infrequent along Pilgrim Creek.
2. **A. drummondii** Gray. Frequent on moraines and outwash plain of the valley.
3. **A. glabra** (L.) Bernh. Frequent in the vicinity of the Oxbow Bend of the Snake River.
4. **A. hirsuta** Scop. Frequent in the meadows east of the Oxbow Bend.
5. **A. holboellii** Hornem. Common on the moraines and sagebrush flats.

   Cauline leaves not auriculate .... var. **pendulocarpa** (A. Nels.) Rollins.

   Cauline leaves auriculate, often clasping ... var. **retrofracta** (Grah.) Rydb.
6. **A. lyallii** S. Wats. Frequent in subalpine and alpine areas.
7. **A. nuttallii** Robinson. Frequent in meadows east of Oxbow Bend of the Snake River.

BARBAREA Winter Cress

Glabrous biennial or perennial herbs. Lvs alt, pinnatifid, and usually somewhat auriculate. Fls borne in racemes; petals yellow; silique linear, 4-angled, beakless; seeds in one row in each chamber.

1. **B. orthoceras** Ledeb. var. **orthoceras.** Infrequent in areas of Oxbow Bend and Colter Bay.

BRASSICA Mustard

Annual to perennial herbs. Lvs alt, variously shaped. Fls

borne in racemes, conspicuous; petals usually yellow to white; siliques linear, circular in cross-section or 4-sided; seeds in one row in each chamber.

†**1. B. campestris** L. Locally common in disturbed sites.

CAPSELLA Shepherd's Purse

Annual herbs. Lvs mostly in a basal rosette. Fls borne in racemes, inconspicuous; petals white; silicles shallowly notched at the apex, compressed contrary to the partition; seeds numerous.

†**1. C. bursa-pastoris** (L.) Medic. Common in disturbed sites throughout the valley.

CARDAMINE Bitter Cress

Mostly glabrous annual to perennial herbs. Lvs simple to pinnate, basal and cauline. Fls borne in racemes; petals white to pink or rose; siliques linear, compressed, usually opening explosively from the base; seeds in a single row in each chamber.

1 Plants annual or biennial from a taproot; petals 2-4 mm long .... **C. oligosperma**
1 Plants perennial with slender rhizomes; petals 3-7 mm long .... **C. breweri**

1. C. breweri Wats. Frequent in wet places along the Snake River.
2. C. oligosperma Nutt. Wet places near the Oxbow Bend of the Snake River.

CHORISPORA Chorispora

Branching annual herbs. Lvs simple to pinnatifid. Fls borne in sparse racemes; petals purplish; siliques indehiscent and terete, long beaked, seeds in a single row in each chamber.

†**1. C. tenella** (Pall.) DC. Infrequent in disturbed sites, east side of the valley.

DESCURAINIA Tansy Mustard

Annual herbs (in ours), pubescent with branched hairs or nearly glabrous. Lvs 1-3 pinnately compound. Fls small, borne in racemes; petals yellow to cream; siliques usually terete, beakless; seeds in 1 or 2 rows to a chamber.

1 Leaves, especially the lower ones, 2-3 times pinnate; siliques 15-30 mm long; septum of fruit with 2-3 veins .. **D. sophia**
1 Leaves mostly pinnate; siliques 15 mm long or less; septum of fruit with 1 vein or none
  2 Siliques linear; pedicels and often the leaves with glandular hairs .... **D. richardsonii**
  2 Siliques club-shaped or nearly so; pedicels and leaves lack glandular hairs .... **D. pinnata**

†**1. D. pinnata** (Walt.) Britt. Frequent in disturbed sites throughout the valley. An extremely variable species.

2. D. richardsonii (Sweet) Schulz. Infrequent at lower to middle elevations in the mountains.

†**3. D. sophia** (L.) Webb. Frequent in disturbed sites throughout the valley.

DRABA Whitlow-wort

Annual or perennial herbs. Lvs with entire to dentate margins, subglabrous to densely pubescent. Fls borne in racemes; petals white or yellow, rounded to bifid at the apex; fruit a few-seeded silicle or short silique more or less flattened parallel to the septum; style short, seeds numerous and wingless.

1 Plants annual
  2 Leaves usually all basal, occasionally some cauline leaves, upper surfaces usually glabrous .... **D. crassifolia**
  2 Leaves usually not all basal, many of them cauline, upper surfaces often pubescent
    3 Pedicels 1-5 times as long as the fruit; silicles 2-3 mm broad .... **D. nemorosa**
    3 Pedicels rarely 1.5 times as long as the fruits; silicles 1.5-2.3 mm broad .... **D. stenoloba**
1 Plants biennial or perennial
  4 Stems leafy, the plant usually not caespitose nor matted
    5 Style generally less than 0.2 mm long or lacking
      6 Pedicels mostly shorter than the silicles . **D. crassifolia**
      6 Pedicels mostly equaling or exceeding the silicles .... **D. stenoloba**
    5 Style present, at least 0.2 mm long
      7 Petals white .... **D. lonchocarpa**
      7 Petals yellow

8 Cauline leaves 2-6; basal leaves fleshy; silicles glabrous .... **D. crassa**
8 Cauline leaves usually more than 6; basal leaves not fleshy; silicles often hairy ..... **D. aurea**
4 Stems usually lacking leaves; plants usually caespitose and mat forming
9 Pubescence of the lower leaf surface mostly sessile or short-stalked, often lying parallel to the leaf axis .... **D. oligosperma**
9 Pubescence of the leaves simple to stellate
10 Leaves usually glabrous except for straight simple hairs .... **D. apiculata**
10 Leaves usually pubescent, often ash colored on both surfaces .... **D. ventosa**

1. **D. apiculata** C. L. Hitchc. var. **apiculata.** Rare, east shore of Timberline Lake, 10,400 ft.
2. **D. aurea** Vahl. Infrequent, forested slopes to alpine meadows.
3. **D. crassa** Rydb. Frequent below the summit of the Middle Teton, 12,300 ft.
4. **D. crassifolia** Graham. Infrequent on talus slopes above Amphitheater Lake.
5. **D. lonchocarpa** Rydb. Infrequent on talus slopes and rock crevices above 9,000 ft.
6. **D. nemorosa** L. Frequent on the moraines of the valley.
7. **D. oligosperma** Hook. Frequent on rocky slopes and exposed ridges above 10,000 ft.
8. **D. stenoloba** Ledeb. Common on moist banks or meadows to dry slopes of the valley.
9. **D. ventosa** Gray. Frequent on the summit of Mt. Moran.

ERYSIMUM Wallflower

Annual to perennial herbs; Lvs alt, simple, sparsely to densely hairy. Fls borne in usually crowded racemes; petals pale yellow, orange or purple-tinged, often pubescent on the back; siliques with strongly nerved valves, from compressed to round-quadrangular in cross-section; seeds in one row in each chamber.

1 Petals 4-8 mm long, yellow .... **E. cheiranthoides**
1 Petals 10-16 mm long, yellow to orange or somewhat reddish .... **E. asperum**

1. **E. asperum** (Nutt.) DC. [*E. capitatum* (Dougl.) Greene].

A highly variable species; infrequent in the lower portions of the canyons.

2. **E. cheiranthoides** L. Infrequent in aspen groves in the area of Signal Mt.

LEPIDIUM Pepper Grass

Annual or perennial herbs. Lvs entire to 2-3 times pinnatifid, sometimes sessile and auriculate. Fls borne in racemes; petals white or yellow or lacking; stamens 2, 4, or 6; silicles ovate to obovate, usually flattened contrary to the partition, notched at the apex; one seed per chamber.

1 Cauline leaves never auriculate or perfoliate
  2 Silicle not notched at apex, usually partially pilose; plants perennial from rhizomes .... **L. latifolium**
  2 Silicle notched at the apex, usually glabrous, annual or biennial, with a tap-root .... **L. densiflorum**
1 Cauline leaves auriculate to almost perfoliate
  3 Petals white; cauline leaves lanceolate to oblong lanceolate .... **L. campestre**
  3 Petals yellow; cauline leaves ovate .... **L. perfoliatum**

†1. **L. campestre** (L.) R. Br. Pepperweed. Common in waste places.

2. **L. densiflorum** Schrad.
   Average silicle about 2.5 mm long; cauline leaves mostly toothed .... var. **densiflorum**
   Average silicle about 3 mm long; cauline leaves entire to subentire .... var. **pubicarpum** (Nels.) Thell.

   Common in the sagebrush flats of the valley.

†3. **L. latifolium** L. Eurasian weed, first seen at Snake River Campground in 1974.

4. **L. perfoliatum** L. Frequent in waste places.

LESQUERELLA Bladder-pod

Low annuals or perennials with dense stellate pubescence. Lvs entire to toothed, nonauriculate. Fls borne in racemes; petals commonly yellow; silicles often inflated, ovate, elliptic, oblong; style persistent; usually several seeds in each chamber. This genus is closely related to *Physaria*.

1 Silicles not flattened or if so, flattened parallel to the partition .... **L. prostrata**
1 Silicles flattened at right angles to the partition
  2 Silicles with strongly flattened margins and strong keels on the flattened sides .... **L. carinata**
  2 Silicles not flattened on the margins or keels lacking .... **L. paysonii**

1. **L. carinata** Rollins. Infrequent in hills above the Gros Ventre River.
2. **L. paysonii** Rollins. Infrequent at the mouth of Pacific Creek.
3. **L. prostrata** Nels. Common along the Snake River and barren hill north of Kelly Warm Springs.

NASTURTIUM Watercress

Annual to perennial glabrous herbs, stems tending to be succulent. Lvs mostly pinnate or pinnatifid. Fls in terminal racemes; petals white, nearly twice the length of the sepals; siliques linear to cylindric; seeds usually in 2 rows in each locule.

1. **N. officinale** R. Br. [*Rorippa nasturtium — aquatcum* (L.) Schinz. & Thell.]. Frequent on delta at the mouth of Cascade Creek, west shore of Jenny Lake.

PHYSARIA Double Bladder-pod

Perennial caespitose herbs with stellate pubescence. Lvs mostly basal, petiolate, nonauriculate. Fls borne in nonbracteate racemes; petals usually yellow; silicles more or less inflated and didymous with a deep apical sinus; style persistent; seeds 2-8.

1. **P. australis** (Pays.) Rollins. Infrequent on switchbacks north of Schoolroom Glacier, also on east boundary of park, road to Gros Ventre Slide.

RORIPPA Yellowcress

Annual to perennial herb growing in damp habitats. Lvs simple to pinnate. Fls borne in nonbracteate racemes; petals yellow scarcely exceeding the sepals; fruit a broad silique, mostly round in cross-section; style short; seeds numerous.

1 Pedicels 4-12 mm long; stems mostly erect, 3-10 mm tall .... **R. palustris**

1 Pedicels 2-4 mm long; stems spreading to decumbent .. .. **R. curvisiliqua**

1. R. curvisiliqua (Hook.) Bess. Locally frequent around some ponds of the valley.

2. R. palustris (L.) Besser. Locally frequent in wet areas in vicinity of the Oxbow Bend of the Snake River.

SMELOWSKIA Smelowskia

Low perennials with branching caudex. Lvs entire to pinnate or pinnatifid. Fls borne in racemes; petals white, cream or purple; siliques short, ovate to linear; mostly somewhat flattened contrary to the partition; stigma expanded, discoid; seeds several.

1. S. calycina C. A. Meyer var. **americana** (Regal & Herd.) Drury & Rollins. Alpine Smelowskia. Frequent on major peaks of the range, above 10,000 ft.

THLASPI Penny-cress

Annual or perennial, glabrous herbs. Lvs simple, entire to lobed, cauline lvs auriculate. Fls borne in terminal racemes; petals white (in ours); silicles obcordate, orbicular, or cuneate, flattened contrary to the partition, thin margined or winged; seeds 2 or several in each chamber.

1 Plants annual; silicles broadly winged and deeply notched at the apex .... **T. arvense**
1 Plants perennial; silicles only slightly winged, truncate or only slightly notched at apex
  2 Styles 0.3-0.5 mm long; petals 2-3.5 mm long, erect .... **T. parviflorum**
  2 Styles 0.6-3.5 mm long; petals 3.6-11.4 mm long, abruptly flared from base of petal .... **T. montanum**

†**1. T. arvense** L. Common in disturbed sites throughout the valley.

2. T. montanum L. var. **montanum.** Frequent on moist or dry, open rocky slopes, 6,000-8,500 ft.

3. T. parviflorum Nels. Infrequent ,often with sagebrush, moist to dry meadows of the valley.

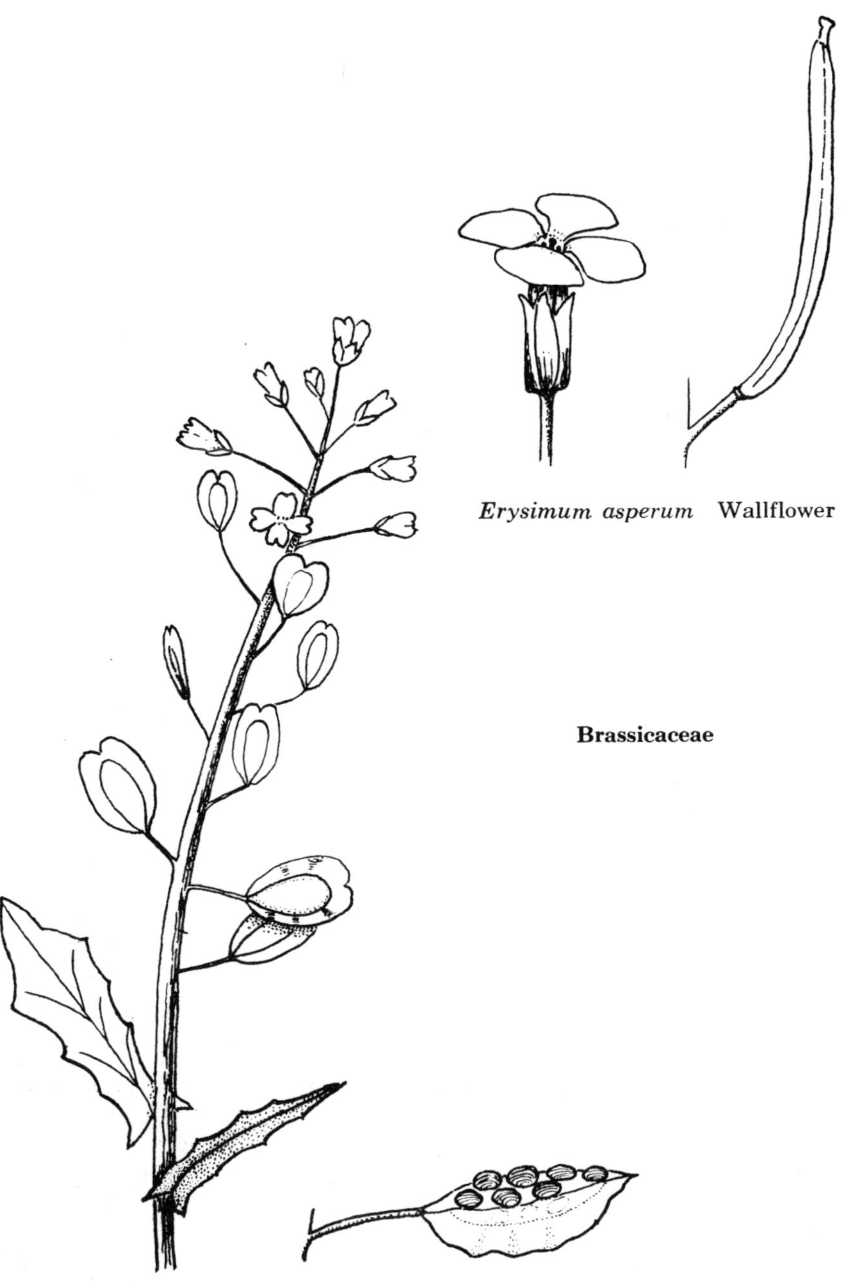

Erysimum asperum Wallflower

Brassicaceae

Thlaspi arvense Penny-cress

CACTACEAE Cactus Family

Stems thick, succulent and spiny. Lvs usually absent. Fls usually solitary and bisex, reg, ov inferior. Ca ∞, Co ∞, S ∞, P 3-∞*. Dry or fleshy berry.

OPUNTIA Prickly-pear Cactus

Fleshy herbs with jointed stems, spiny. Fls in areoles of previous year's stem; sepals green, but grading into the yellow to red petals; berry dry in our spp.

1 Stem joints distinctly flattened, firmly attached; flowers yellow to reddish .... **O. polyacantha**
1 Stem joints not flattened, readily separated from plant; flowers yellow .... **O. fragilis**

1. **O. fragilis** (Nutt.) Haw. Frequent on gravel banks of the Snake River in area of Deadman's Bar also at Kelly Warm Springs.
2. **O. polyacantha** Haw. Not seen in G.T.N.P. but frequent in Y.N.P. especially in Mammoth Area.

CALLITRICHACEAE Water-starwort Family

Aquatic herbs. Lvs opp, simple, no stip. Fls solit, unisex, reg, ov superior. Per 0, S 1, P 2*, deeply 4-lobed, splitting at maturity into 4 1-seeded fruits.

CALLITRICHE Water-starwort

Characteristics of the family.

1. **C. verna** L. (*C. palustris* of many authors). Common in pools and adjacent mud along Snake River.

CAMPANULACEAE Bellflower Family

Herbs, annual or perennial. Lvs usually alt, simple, no stip. Infl various. Fls bisex, reg, ov inferior. Ca 5*, Co 5*, S 5, P 2-5*. Capsule.

1 Corolla irregular; filaments and anthers united into a tube .... **Porterella**
1 Corolla regular; filaments and anthers distinct

2 Plants annual; lower flowers cleistogamous, often with reduced corolla .... **Heterocodon**
2 Plants perennial; no cleistogamy, well developed corolla .... **Campanula**

CAMPANULA Harebell

Chiefly perennial herbs. Lvs alt, often dimorphic, from linear to ovate. Fls borne in a racemiform inflorescence; corolla blue or violet, more or less campanulate; stamens 5, free from the corolla; stigma 3-5; capsule trilocular opening by pores.

1. **C. rotundifolia** L. Frequent in a variety of habitats from disturbed roadside to coniferous forest.

HETEROCODON Heterocodon

Annual, slender herbs. Lvs alt, sessile and toothed. Fls in lax false spikes; lower fls cleistogamous with reduced or abortive corollas; upper fls with normal corollas; stamens separate from corolla; ov inferior with 3 locules; stigma 3-lobed. Capsule.

1. **H. rariflorum** Nutt. Reported in Teton Canyon.

PORTERELLA Porterella

Succulent annuals. Lvs alt, sessile. Fls axillary, pedicellate; calyx lobes linear; corolla bilabiate, blue to purple with a yellow throat; filaments and anthers connate; fruit a capsule.

1. **P. carnosula** (H. & A.) Torr. Locally common at the margins of drying ponds, from Signal Mt. north.

CAPRIFOLIACEAE Honeysuckle Family

Woody shrubs. Lvs opp, simple or compound, no stip. Infl often cymose. Fls bisex, reg/irreg, ov inferior. Ca 4-5*, Co 4-5*, S 4-5, P 3-5*. Berry.

1 Plants creeping, evergreen, herblike shrubs .... **Linnaea**
1 Plants erect, deciduous shrubs
  2 Style very short or absent; inflorescence branched and generally with many flowers .... **Sambucus**
  2 Style well developed and more or less elongated; inflorescence usually of paired flowers

Campanula rotundifolia Harebell **Campanulaceae**

3 Corolla spurred at base; fruit red or purplish black . . . . **Lonicera**

3 Corolla with only a slight swelling at the side; fruit white . . . . **Symphoricarpos**

LINNAEA Twinflower

Evergreen, trailing subshrub. Lvs opp, entire to crenate. Fls mostly paired on terminal peduncles, 5-merous; corolla pink or pinkish; stamens 4; fruit small, dry, indehiscent.

1. **L. borealis** L. ssp. **longiflora** (Torr.) Hult. Locally frequent in shade of lodgepole pine; slopes of Signal Mt.

LONICERA Honeysuckle

Erect shrubs or woody vines. Lvs opp, usually entire. Fls borne in spikes or in pairs; calyx very small, 5 toothed, sometimes lacking; corolla commonly bilabiate, yellow (in ours), the tube often gibbous or spurred near the base; stamens 5; stigma capitate; fruit a berry.

1 Bracts at the end of the peduncle foliaceous; berries of the two flowers distinct, subtended by a colored involucre . . . . **L. involucrata**

1 Bracts minute, not foliaceous; berries of the two flowers more or less united; involucre absent . . . . **L. utahensis**

1. **L. involucrata** (Rich.) Banks. Bearberry Honeysuckle. Frequent around the shores of the valley lakes.
2. **L. utahensis** S. Wats. Utah Honeysuckle. Common in the open coniferous forest of the valley and canyons.

SAMBUCUS Elderberry

Shrubs with pithy stems. Lvs large, pinnately compound, serrate leaflets. Fls borne in compound cymes; calyx lobes minute; corolla nearly rotate, 3-5 lobes, whitish; stamens 5; fruit berry-like.

1. **S. racemosa** L. ssp. **pubens** (Michx.) House.

Fruit black or purplish black . . . . var. **melanocarpa** (Gray) McMinn.

Fruit bright red . . . . var. **microbotrys** (Rydb.) Kearney & Peebles.

Frequent around the shores of the valley lakes and extending into major canyons.

SYMPHORICARPOS Snowberry

Erect or trailing shrubs. Lvs opp, exstipulate, mostly entire. Fls borne in terminal or axillary clusters; usually 5-merous as to perianth and stamens; corolla pink to white; fruit berrylike, white.

1 Corolla tubular or funnel-shaped, longer than wide, the lobes mostly ¼-½ as long as the tube .... **S. oreophilus**
1 Corolla short-campanulate, not usually longer than wide, the lobes from ½ as long to longer than the tube
  2 Style and stamens exserted; style long, hairy near the middle .... **S. occidentalis**
  2 Style and stamens shorter than or only equaling the corolla, not exserted .... **S. albus**

1. **S. albus** (L.) Blake var. **laevigatus** (Fern.) Blake. Frequent in sagebrush flats and on moraines of the valley.
2. **S. occidentalis** Hook. Frequent in open aspen stands of the valley.
3. **S. oreophilus** Gray var. **utahensis** (Rydb.) A. Nels. Frequent in sagebrush flats and moraines of the valley. [*S. vaccinioides* Rydb., *S. tetonensis* A. Nels., *S. rotundifolius* Gray].

CARYOPHYLLACEAE Pink Family

Herbs, annual or perennial. Lvs usually opp, entire, usually no stip. Infl cymose/fls solitary. Fls usually bisex, reg, ov superior. Ca 4-5*, Co 4-5, S 3-10, P 2-5*, free central placentation. Capsule.

1 Leaves with obvious non-green stipules .... **Spergularia**
1 Leaves without stipules
  2 Sepals separate and distinct to the base
    3 Styles usually 5; capsule cylindric, opening with 10 teeth at apex .... **Cerastium**
    3 Styles usually 3, 4 or 5; capsule ovate or oblong, opening by fewer than 10 teeth
      4 Styles 4 or 5, alternating with the sepals ...... **Sagina**
      4 Styles usually 3
        5 Petals 2-parted or lacking; stamens and petals inserted under the ovary (use hand lens) .. .. **Stellaria**

5 Petals not 2-parted or sometimes lacking; stamens and petals often inserted at the edge of a distinct disk .... **Arenaria**
2 Sepals united, forming a distinct tube
6 Calyx lobes conspicuous and foliaceous; flowers solitary on axillary peduncles .... **Agrostemma**
6 Calyx lobes short and not foliaceous; flowers not solitary on axillary peduncles
7 Styles commonly 2
8 Petals bearing appendages at the union of the blade and the claw .... **Saponaria**
8 Petals not bearing appendages .... **Vaccaria**
7 Styles commonly 3 to 5
9 Styles usually 3 .... **Silene**
9 Styles usually 5 .... **Lychnis**

AGROSTEMMA Corncockle

Annual herbs. Lvs opp, narrow, exstipulate. Fls several in open cymes; sepals united, but the calyx lobes longer than the tube; petals reds (in ours) about twice as long as calyx tube; capsule; seeds black.

†**1. A. githago** L. European weed of disturbed sites.

ARENARIA Sandwort

Annual to perennial herbs. Lvs linear to broad, opp, exstipulate. Fls several in diffuse to capitate cymes, but occasionally single; sepals 5; petals 5, entire or slightly emarginate, usually white; stamens 10; styles usually 3; capsule 1-loculed.

1 Leaves narrowly oblong to ovate, not sharp pointed .... **A. lateriflora**
1 Leaves mostly narrowly linear and sharp pointed
2 Plants with flowering stems over 10 cm tall; capsule dehiscing by 6 valves .... **A. congesta**
2 Plants mostly low and matted, usually not over 10 cm tall; capsule dehiscing by 3 valves
3 Sepals oblong, rounded at apex; petals longer than sepals .... **A. obtusiloba**
3 Sepals lanceolate, acute at apex; petals equaling or shorter than sepals .... **A. nuttallii**

1. **A. congesta** Nutt. Common in the sagebrush community; also at Timberline Lake.
2. **A. lateriflora** L. Infrequent in moist shaded areas of the valley.
3. **A. nuttallii** Pax. Rare on divide between So. Cascade and Alaska Basin.
4. **A. obtusiloba** (Rydb.) Fern. Frequent on major peaks above timberline.

CERASTIUM Chickweed; Cerastium

Annual, to perennial herbs, often forming roots at the nodes. Lvs opp, exstipulate. Fls in terminal cymes; sepals 5, distinct to the base; petals 5, white, 2 lobed or 2 cleft; stamens usually 10; styles usually 5; capsule cylindric, opening usually by 10 teeth.

1 Petals about equal to the sepals . . . . **C. vulgatum**
1 Petals at least 1.5 times longer than the sepals
  2 Bracts below the flowers with a scarious margin; up to subalpine . . . . **C. arvense**
  2 Bracts below the flowers rarely scarious; tundra and alpine . . . . **C. beeringianum**

1. **C. arvense** L. Frequent, dry to moist meadows along the Snake River.
2. **C. beeringianum** Cham. & Schlecht. Frequent above timberline on Skyline Trail.

†3. **C. vulgatum** L. A Eurasian weed reported in Teton Canyon and Y.N.P.

LYCHNIS Campion

Annual to perennial herbs. Lvs opp, entire. Fls perfect or imperfect, cymose; sepals united into a tube, often inflated in fruit; petals white to deep red, with a narrow claw; stamens 10; styles usually 5; capsule dehiscent by 5 or 10 teeth.

1 Flowers white, imperfect, plants dioecious; blades of the petals over 5 mm long . . . . **L. alba**
1 Flowers white or pinkish, mostly perfect; blade of the petals less than 5 mm long . . . . **L. drummondii**

†1. **L. alba** Mill. A European species now widely distributed in N. America.

2. **L. drummondii** (Hook.) Wats. Frequent in sagebrush, meadows, and open woods in Teton Canyon, 11 miles east of Driggs, Idaho.

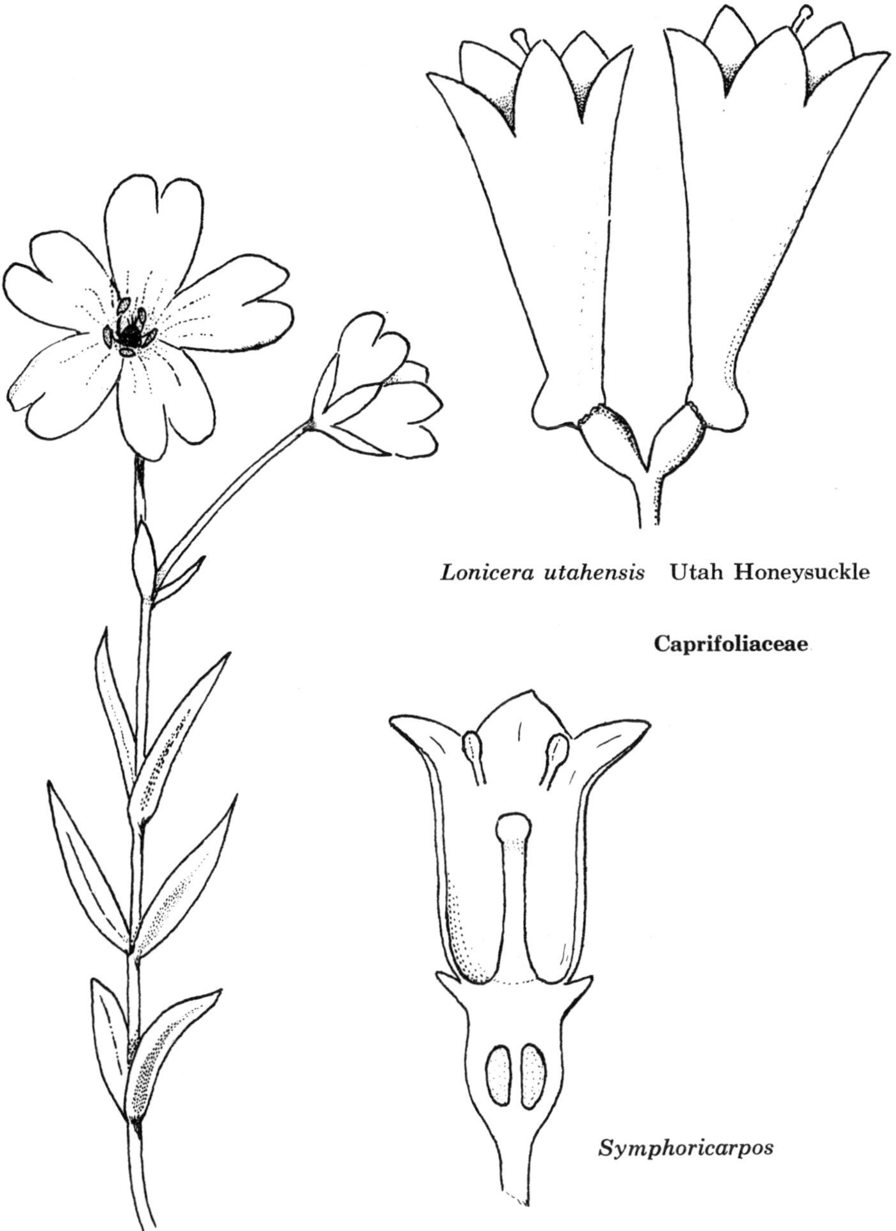

Lonicera utahensis Utah Honeysuckle

Caprifoliaceae

Symphoricarpos

Cerastium arvense Chickweed

Caryophyllaceae

SAGINA Pearlwort

Annual or perennial herbs often matted. Lvs opp, linear. Fls inconspicuous, axillary to cymose; sepals 4 or 5, distinct; petals 4 or 5, white; styles usually equal in number to and alternate with the sepals, capsule valves as many as the sepals and opposite them.

1. **S. saginoides** (L.) Britt. Common in moist areas, especially on mud flats, circumpolar.

SAPONARIA Bouncing Bett; Soapwort

Perennial herbs with rhizomes. Lvs lanceolate to oblanceolate. Fls cymose, complete; sepals connate; petals clawed, appendaged at top of the claw, white to pink, sometimes doubled; stamens 10; styles usually 2; capsule 1 loculed.

1. **S. officinalis** L. Rare, European ornamental established behind Park Hdqts. Bld.

SILENE Campion; Catchfly; Wild Pink

Annual to perennial herbs. Lvs opp or whorled. Fls cymose, perfect to imperfect; sepals united into a tube, often inflated in fruit; petals 5, white to pink, red or purple, blade entire to lobed; stamens 10; styles usually 3; capsule dehiscent by 6-10 teeth.

1 Plants annual, usually in disturbed sites
  2 Upper internodes glabrous, often glandular in bands just below the nodes; blade of petals 2-4 mm long .... **S. antirrhina**
  2 Upper internodes pubescent, not glandular in bands; blade of petals 6-10 mm long .... **S. noctiflora**
1 Plants perennial, usually on undisturbed soil
  3 Plants cushion-like, less than 10 cm tall .... **S. acaulis**
  3 Plants not cushion-like, usually over 10 cm tall
    4 Petals generally less than 10 mm long; flowers in open leafy cymes .... **S. menziesii**
    4 Petals generally more than 12 mm long; flowers in reduced inflorescences with bracts
      5 Petals with 4-6 appendages .... **S. oregana**
      5 Petals with 2 appendages
        6 Calyx purplish; capsule 3-4 loculed at least in the lower portion .... **S. repens**
        6 Calyx seldom purplish; capsule 1-loculed . **S. parryi**

1. **S. antirrhina** L. A weedy species apparently native to No. Am., reported in Y.N.P.
2. **S. acaulis** L. Frequent in rock crevices above 9,500 ft.
3. **S. menziesii** Hook. var. **menziesii.** Infrequent in moist shaded areas of the valley.

†4. **S. noctiflora** L. Infrequent weed of disturbed sites.

5. **S. oregana** Wats. Frequent on subalpine slopes in Teton Canyon.
6. **S. parryi** (Wats.) Hitchc. & Mag. Infrequent in canyons above 9,500 ft.
7. **S. repens** Pers. Reported in Hoback Canyon.

SPERGULARIA Sandspurry

Annual or perennial herbs. Lvs often with scarious stipules. Fls axillary or in leafy cymes; sepals 5 distinct; petals 5, white to pink; stamens 2-10; styles 3; capsule dehiscing by 3 valves.

†1. **S. rubra** (L.) Presl. Common weed of disturbed sites; native of Europe.

STELLARIA Starwort

Slender annual or perennial herbs, often with rhizomes. Fls axillary to cymose; sepals 5, distinct; petals 5, white, 2-cleft or parted; stamens 3 to 10; styles usually 3; capsule dehiscing by twice as many teeth as styles. A very confusing and variable group of species.

1 Leaves tending to be oblong or ovate; flowers often single in leaf axils; sepals usually longer than the petals
  2 Leaf margins crisped; sepals acute .... **S. crispa**
  2 Leaf margins not crisped; sepals obtuse .... **S. obtusa**
1 Leaves linear to lanceolate; flowers usually in cymes or umbels; sepals often shorter than the petals
  3 Petals minute or absent; flowers in umbellate cymes .... **S. umbellata**
  3 Petals usually longer than the sepals; flowers in cymes or solitary
    4 Flowers axillary not cymose; petals and sepals about equal
      5 Leaves mostly linear or linear-lanceolate, sharply acute and stiff .... **S. longipes**

5 Leaves linear-lanceolate to oblong, not stiff .... **S. crassifolia**

4 Flowers few to many and cymose; petals sometimes less than sepals

6 Cymes with leafy bracts; petals less than the sepals, or lacking .... **S. calycantha**

6 Cymes with membranous bracts, or petals longer than sepals .... **S. longifolia**

1. **S. calycantha** (Ledeb.) Bong. Major canyons of the range.
2. **S. crassifolia** Ehrb. Moist areas; reported in Y.N.P.
3. **S. crispa** Cham. & Schlecht. Infrequent in moist places, ½ mile above White Grass Ranch.
4. **S. longifolia** Muhl. Frequent along the Snake River.
5. **S. longipes** Goldie. Common in the meadows near the Oxbow Bend.
6. **S. obtusa** Engelm. Mountain meadows and streambanks. Reported in Y.N.P.
7. **S. umbellata** Turcz. Infrequent in South Teton Canyon near Skyline Trail.

VACCARIA Cowherb; Cockle

Annual herbs. Lvs opp. Fls cymose; sepals united, 10-nerved and winged; petals pink, clawed but no appendages; styles 2; capsule 1-loculed.

†1. **V. pyramidata** Medic. European weed of roadsides. Reported in Y.N.P.

CELASTRACEAE Stafftree Family

Woody shrubs. Lvs alt/opposite, simple, stip/no stip. Infl cymes. Fls usually bisex, reg, ov superior. Ca 4-5*, Co 4-5, S 4-5, P 2-5*. Capsule or follicle.

PACHISTIMA Mountain-lover

Evergreen shrubs. Lvs opp, leathery. Fls axillary; 4-merous except for 2 carpels in ovary; capsule.

1. **P. myrsinites** (Pursh) Raf. Common on the moraines and shores of the valley lakes and extending up to 9,000 ft. in the mountains.

CERATOPHYLLACEAE Hornwort Family

Aquatic herbs. Lvs whorled, much divided, no stip. Fls solitary, unisex, reg, ov superior. Per 0, 8-12 cleft involucre, S ∞, P 1. Nut or achene.

CERATOPHYLLUM Hornwort

Characteristics of the family.

1. **C. demersum** L. Frequent in lakes, ponds, and slow moving streams of the valley.

CHENOPODIACEAE Goosefoot Family

Herbs/shrubs. Lvs alt/opp, simple, no stip/reduced to scales. Infl usually cymose. Fls uni/bisex, reg, ov superior. Ca 1-5 lobed, Co 0, S 2-5, P 2-3*. Utricle.

1 Shrubs with spinose branches; leaves linear and semiterete; embryo spirally coiled .... **Sarcobatus**
1 Shrubs or herbs, if shrubby and spiny, then leaves flattened; embryo annular or sometimes spiral
  2 Plants copiously white pubescent, mostly stellate, monoecious shrubs with narrow alternate leaves . **Ceratoides**
  2 Plants various but not as above
    3 Leaves awl-shaped, spine-tipped at maturity, a tumbleweed of disturbed sites .... **Salsola**
    3 Leaves not spine-tipped except for a few hair-like bristles
      4 Flowers all axillary, borne singly or in clusters of 2-5 along main stem; leaves glabrous, linear often nearly terete .... **Suaeda**
      4 Fls usually in crowded spikes or panicles; leaves various in shape but not as above
        5 Perianth lacking or consisting of a single bract-like tepal; annual; leaves often with hastate bases .... **Monolepis**
        5 Perianth 3-5 lobed; annual or perennial; leaves often toothed
          6 Plants monoecious, male flowers with 3-5 lobed perianth .... **Atriplex**
          6 Plants with mostly perfect flowers and usually a lobed perianth

7 Perianth not winged in fruit .. **Chenopodium**
7 Perianth often becoming winged or keeled in fruit .... **Kochia**

ATRIPLEX Greasewood; Shadscale; Saltbush

Annual or perennial herbs or shrubs, sometimes spiny, glabrous to covered with a granular powder. Plants monoecious or dioecious, single or in clusters; male flowers with no bracts, perianth usually with 5 parts; stamens usually adnate to perianth; stigmas 2-3; embryo annular. This genus is closely related to *Chenopodium* and species of these genera are often confused.

†**1. A. patula** L. var. **hastata** (L.) Gray. Widespread weed in disturbed sites.

CHENOPODIUM Lamb's Quarter; Pigweed; Goosefoot

Annual or perennial herbs, glabrous to powder covered. Lvs alt, more or less fleshy, entire, hastate or toothed, often becoming reddish. Fls perfect, greenish usually in clusters, inflorescences freely branched.

1 Flowers densely compacted in clusters or heads; perianth often reddish at maturity; fruit laterally flattened
2 Compact head-like inflorescences less than 4 mm broad; perianth not reddish or fleshy .... **C. rubrum**
2 Compact head-like inflorescences often greater than 4 mm broad at maturity; perianth often reddish and fleshy
3 Perianth becoming reddish and fleshy as fruits develop .... **C. capitatum**
3 Perianth greenish to somewhat pinkish, but not fleshy as fruits develop .... **C.chenopodioides**
1 Flowers not all in densely compacted clusters or heads; perianth not reddish at maturity; fruit flattened at top
4 Leaf blades mostly toothed on margins and often hastate; fruit wall usually fused tightly to the seed .. **C. album**
4 Leaf blades entire, often with large hastate lobes; fruit wall not tightly fused to seed .... **C. atrovirens**

†**1. C. album** L. Locally common in waste places.
†**2. C. atrovirens** Rydb. Disturbed sites near Jackson Lake Dam.
†**3. C. capitatum** (L.) Asch. Locally common in waste places.

Chenopdium album Lamb's Quarter

Chenopodiaceae

Celastraceae

Pachistima myrsinites Mountain-lover

†4. **C. chenopodioides** (L.) Aellen. Roadsire weed in the valley.
†5. **C. rubrum** L. Widely distributed weed throughout Teton Co.

CERATOIDES Winterfat; Winter Sage

Small, perennial shrubs. Lvs alt, with densely stellate pubescence. Plants monoecious, perianth of 4 parts, female flowers naked but surrounded by 2 villous, slightly keeled bracteoles.

1. **C. lanata** (Pursh) Howell [*Eurotia lanata* (Pursh) Moq.]. Frequent in saline areas east of Elk Ranch Reservoir.

KOCHIA Summer Cypress

Annual herbs (in ours) to shrubby perennials. Lvs alt, linear to narrowly lanceolate. Fls sessile in axils of foliose bracts, usually perfect; perianth of 5 parts and surrounding the fruit, often becoming keeled and winged.

†1. **K. scoparia** (L.) Schrad. Common in waste places of the valley.

MONOLEPIS Povertyweed

Annual herbs, covered with granular-like powder. Lvs alt, entire to hastate, fleshy. Fls perfect to plants monoecious in axillary clusters; perianth usually a single bract-like tepal; stamens 1-2; fruit more or less flattened.

†1. **M. nuttalliana** (Schultes) Greene. Common in disturbed sites of the valley.

SALSOLA Russian Thistle; Saltwort

Annual (in ours) or perennial herbs. Lvs alt, linear, to awl-shaped, usually with a spine tip at maturity. Fls perfect, solitary or in clusters; perianth usually 3 and distinct to base.

†1. **S. kali** L. Ubiquitous Eurasian weed of disturbed sites of the valley.

SARCOBATUS Greasewood

Branched, spiny, perennial shrubs. Lvs alt, fleshy, and linear. Fls imperfect, plants monoecious, many flowered spikes; ovary surrounded by a cup-like perianth which becomes enlarged in fruit.

1. **S. vermiculatus** (Hook.) Torr. Frequent in saline areas east of Elk Ranch Reservoir.

SUAEDA Seablite; Seepweed

Annual (in ours) or perennial herbs or small shrubs. Lvs alt, linear, terete or more or less flattened. Fls perfect to imperfect, plants sometimes monoecious, fls in axils of small bracts; perianth fleshy, lobed about ½ the length, more or less hood-shaped.

1. **S. depressa** (Pursh) Wats. Frequent in saline areas 2 miles east of Elk Ranch Reservoir.

CONVOLVULACEAE Morning-glory Family

Herbs (in ours), sometimes with milky juice. Lvs alt, simple, no stip. Infl cymose/fls solitary. Fls usually bisex, reg, ov superior. Ca 5, Co 5*, S 5, P 2*. Capsule.

CONVOLVULUS Bindweed; Morning-glory

Perennial herbs from slender rhizomes, trailing or twining stems. Lvs alt, simple, more or less hastate. Fls funnel-shaped, single or paired on axillary peduncles; sepals 5; corolla 5-lobed, white, pinkish or purple; stamens 5; ovary 2-loculed; capsule with 2-4 valves.

†**1. C. arvensis** L. European weed, warm spring, north of Kelly.

CORNACEAE Dogwood Family

Woody. Lvs alt/opp, simple no stip. Infl cymose. Fls uni/bisex, reg, ov inferior. Ca 4-5, Co 4-5, S 4-5, P 2*. Drupe/berry.

CORNUS Dogwood

Trailing subshrub to shrub (in ours). Lvs with obvious pinnate veins. Fls subtended by 4-8 white or pinkish petal-like bracts or bracts lacking; petals white or greenish; stamens 4; drupe 2 seeded.

1 Plants woody shrub; leaves opposite; inflorescence not prominently bracteate .... **C. stolonifera**

1 Plants low trailing subshrub; leaves whorled at the top; flowers subtended by 4 white bracts .... **C. canadensis**

1. C. canadensis L. Not seen in G.T.N.P. but frequent in Y.N.P. in moist woods.

2. C. stolonifera Michx. Redosier Dogwood. Frequent at Park Hdqts. and the shores of the valley lakes.

CRASSULACEAE Stonecrop Family

Succulent herbs. Lvs opp/alt, simple, no stip. Infl cymose. Fls bisex, reg, ov superior. Ca 4-5, Co 4-5*, S 8-10, P 4-5. Follicle.

SEDUM Stonecrop

Succulent perennial herbs. Lvs alt or opp, fleshy, terete or flat. Fls perfect or sometimes imperfect; calyx 4-5 parted; petals as many as sepals, distinct or united at the base, usually yellow, pink or purple; stamens usually twice as many as the petals; carpels 4-5; fruit a follicle.

1 Flowers usually purple or pink; leaves flat, mostly on flowering stem and persistent through the season
  2 Flowers mostly imperfect; petals usually purple, 5 mm long or less .... **S. rosea**
  2 Flowers perfect; petals pinkish, mostly 8-18 mm long .... **S. rhodanthum**
1 Flowers usually yellow; leaves mostly basal along trailing stems; leaves of flowering stem usually deciduous by the time of flowering
  3 Leaves opposite, under 1 cm long, broadly oval to obovate .... **S. debile**
  3 Leaves alternate, of various shapes; often over 1 cm long
    4 Leaves strongly keeled, linear or linear lanceolate, strongly acuminate; follicles widely divergent at maturity .... **S. stenopetalum**
    4 Leaves not strongly keeled, more or less terete; follicles erect .... **S. lanceolatum**

1. S. debile Wats. Frequent on rocky ledges along the lower slopes of the range.

2. S. lanceolatum Torr. Common in the outwash plain of the

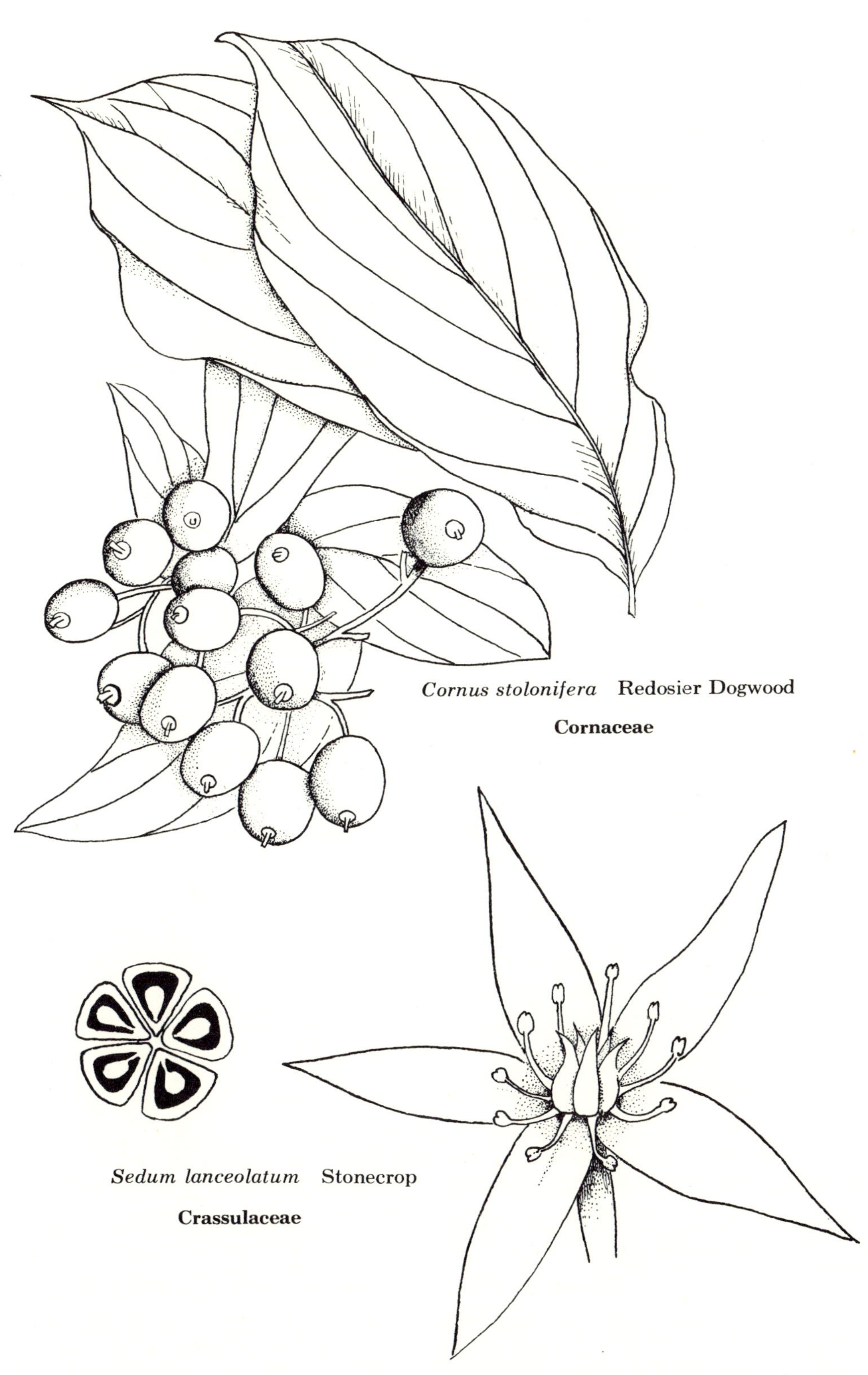

Cornus stolonifera Redosier Dogwood

Cornaceae

Sedum lanceolatum Stonecrop

Crassulaceae

valley. This species is usually recognized under the name *S. stenopetalum* Pursh. See Hitchcock et. al. 1964 pp. 569.

3. **S. rhodanthum** Gray. Frequent along streams in Indian Paintbrush and Snowshoe Canyons above 8,000 ft.
4. **S. rosea** (L.) Scop. ssp. **integrifolium** (Raf.) Hult. Frequent in moist meadows and stream banks above 9,000 ft.
5. **S. stenopetalum** Pursh. Dry gravelly moraines of the valley [*S. douglasii* Hook.].

DROSERACEAE Sundew Family

Herbs. Lvs in rosettes, usually simple, insectiverous with glandular hairs. Infl raceme/panicle. Fls bisex, reg, ov superior or slightly inferior. Ca 4-5*, Co 5, S 5-20, P 3-5*. Capsule.

DROSERA Sundew

Characteristics of the family.

1. **D. anglica** Huds. Rare, sphagnum site, south shore of Leigh Lake.

ELAEAGNACEAE Oleaster Family

Woody. Lvs alt/opp, simple, no stip, with stellate pubescence. Infl various. Fls uni/bisex, reg, ov superior. Per 4*, S 4-8, P 1. Fruit drupaceous.

1 Plants usually with perfect flowers, stamens 4; Lvs alternate .... **Elaeagnus**
1 Plants dioecious; stamens 8; leaves opposite ...... **Shepherdia**

ELAEAGNUS Elaeagnus

Shrubs: Lvs alt. Fls axillary; calyx lobed less than half the length; stamens 4; ovary wall maturing into a hardened bony drupe.

1. **E. commutata** Bernh. Silverberry. Frequent along the Snake River.

SHEPHERDIA Buffalo-berry

Shrubs. Lvs opp. Fls axillary, unisexual, appearing early in the spring before the leaves; ovary wall not greatly hardened in fruit.

1. **S. canadensis** (L.) Nutt. Common along the Snake River and the shores of the valley lakes.

ERICACEAE Heath Family

Woody/herbs. Lvs alt/opp/basal, simple, no stip. Infl racemose/clustered/fls solit. Fls bisex, reg, ov superior/inferior. Ca 4-5(*), Co 4-5(*), S 5-10 often poricidal, P 2-10*. Capsule/berry/drupe.

1 Ovary inferior or appearing so because of fleshy calyx
  2 Ovary truly inferior; erect shrubs with deciduous leaves .... **Vaccinium**
  2 Ovary superior but surrounded by the fleshy calyx when mature and thus appearing inferior; mostly prostrate evergreen shrubs .... **Gaultheria**
1 Ovary superior, free of the calyx
  3 Fruit fleshy; evergreen prostrate shrubs .... **Arctostaphylos**
  3 Fruit a dry capsule; erect shrubs
    4 Corolla of separate petals .... **Ledum**
    4 Corolla of united petals
      5 Corolla rotate (saucerlike) leaves with revolute margins .... **Kalmia**
      5 Corolla urceolate
        6 Leaves evergreen; plants less than 1 m tall .... **Phyllodoce**
        6 Leaves deciduous; plants usually 1-5 m tall .... **Menziesia**

ARCTOSTAPHYLOS Manzanita

Evergreen shrubs, prostrate and matted (in ours). Lvs alt, leathery, entire. Fls borne in short racemes; sepals 5; corolla urceolate, white or light pink, with 4-5 lobes; stamens 8-10; fruit fleshy, berry-like.

1. **A. uva-ursi** (L.) Spreng. Bearberry, Kinnikinnick. Frequent along the Snake River and around the shores of some valley lakes; also Avalanche Canyon above 10,000 ft.

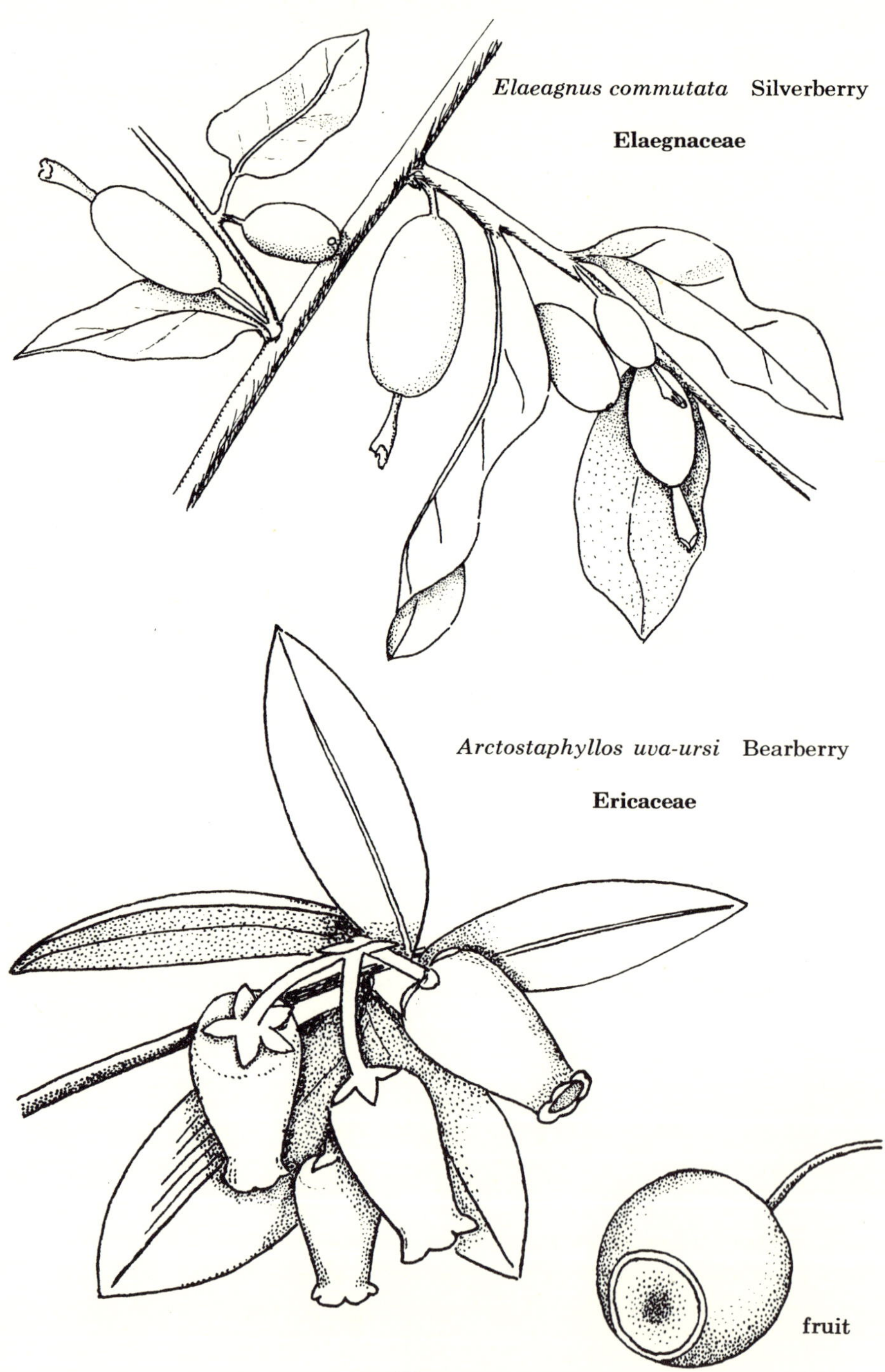
Elaeagnus commutata Silverberry
Elaegnaceae
Arctostaphyllos uva-ursi Bearberry
Ericaceae
fruit
Vaccinium membranaceum Huckleberry

GAULTHERIA Wintergreen

Matted to semierect shrubs. Lvs alt, evergreen, leathery. Fls solitary and axillary; perianth 4-5 merous; corolla white or pinkish, campanulate to urceolate; stamens 8-10; fruit a berry-like capsule.

1. **G. humifusa** (Grah.) Rydb. Western Wintergreen. Infrequent on moist soil on the shores of Bradley, String and Amphitheater Lakes.

KALMIA Swamp Laurel

Low, evergreen shrubs. Lvs alt to whorled, entire and leathery. Fls in clusters of 3-10, on slender red pedicels; calyx lobed nearly to the base; petals united into a shallow bowl-shape, with 10 recessed sacs which hold the anthers in bud, pink to rose-purple; fruit a sepiticidal capsule.

1. **K. microphylla** (Hook.) Heller. Frequent on the shores of the valley lakes and some subalpine lakes.

LEDUM Labrador Tea

Evergreen shrubs. Lvs alt, leathery, revolute leaf margins. Fls in racemes or corymbs; perianth 5-merous; corolla white; fruit a capsule.

1. **L. glandulosum** Nutt. Not collected in G.T.N.P., but known to be within 40 miles of north boundary.

MENZIESIA False Huckleberry

Erect slender shrubs up to 2 m tall. Lvs alt, deciduous, thin. Fls in terminal clusters, on shoots of previous year; calyx very short and lobed; corolla urceolate, 8-9 mm long, pinkish; stamens 8 (in ours); septicidal capsule.

1. **M. ferruginea** Smith var. **glabella** (Gray) Peck. Frequent in moist woods and mountain stream banks up to 8,500 ft.

PHYLLODOCE Mountain Heath; Mountain Heather

Dwarf alpine shrubs. Lvs alt, evergreen, closely crowded, linear. Fls single from the axils, clustered; perianth 5-merous; corolla yellow or pink, campanulate to urceolate; stamens 10; capsule septicidal.

1 Corolla pink, glabrous; sepals obtuse .... **P. empetriformis**
1 Corolla yellow, glandular hairs on the outside; sepals acute .... **P. glanduliflora**

1. **P. empetriformis** (Sw.) D. Don. Frequent around the shores of lakes above 9,000 ft.
2. **P. glanduliflora** (Hook.) Cov. Infrequent on the shores of lakes above 9,000 ft.
3. **Phyllodoce X intermedia** (Hook.) Camp. Hybrid between *P. empetriformis* and *P. glanduliflora.* Occasionally seen at Holly and Amphitheater Lakes.

VACCINIUM Huckleberry; Blueberry

Low shrubs. Lvs alt, deciduous (in ours). Fls one to several on drooping pedicels; perianth 4-5 merous; corolla pink to white, globose to urceolate; stamens 8-10; fruit reddish to bluish berry, juicy and edible.

1 Branches prominently angled, greenish; berries bright red .... **V. scoparium**
1 Branches usually not prominently angled, brownish; berries usually blue to blackish
  2 Calyx strongly lobed, persistent in fruit; 1-4 flowers per axil .... **V. occidentale**
  2 Calyx shallowly lobed, deciduous; single flower per axil
    3 Plants 2-4 dm tall; leaves more or less serrate above the mid-point; berries 5-8 mm in diam. .. **V. caespitosum**
    3 Plants 4-18 dm tall; leaves sometimes entire or serrate below mid-point; berries 6-10 mm in diam .... **V. membranaceum**

1. **V. caespitosum** Michx. Mountain meadows and slopes; near Colter Bay.
2. **V. membranaceum** Dougl. Abundant in the canyons and around the shores of the valley lakes.
3. **V. occidentale** Gray. Infrequent near the outlet of Bradley Lake and small pond east of Colter Bay.
4. **V. scoparium** Leib. Abundant in the canyons and on morainal soil of the valley.

FABACEAE (LEGUMINOSAE) Pea Family

Herbs/woody. Lvs usually alt, compound/simple, stip. Infl

various but usually racemose. Fls bisex, irreg (papilionaceous in ours), ov superior. Ca 5(*), Co 5, S 5-∞, diadelphous, monadelphous or distinct, P 1. Legume.

1 Plants shrubs or small trees .... **Caragana**
1 Plants herbs, sometimes woody at the base
 2 Leaves ending in a tendril .... **Vicia**
 2 Leaves never having a tendril
  3 Leaves dotted with small glands (use hand lens); fruits covered with spines .... **Glycyrrhiza**
  3 Leaves seldom dotted with glands; fruits without spines
   4 Leaves trifoliate; leaflet margins denticulate to serrate (rarely entire in *Trifolium*)
    5 Fruit sickle-shaped or spirally coiled; flowers yellow or purplish .... **Medicago**
    5 Fruit straight or nearly so; flowers often other than yellow or purple
     6 Flowers in long narrow racemes, yellow or white; plants sweet smelling .... **Melilotus**
     6 Flowers in heads or short spikes, white, pink or purple; plants hardly aromatic **Trifolium**
   4 Leaves rarely trifoliate, but if so, then the leaflets entire
    7 Fruits 1-2 seeded and covered with short spiny teeth, nearly as broad as long .... **Onobrychis**
    7 Fruits several seeded and not spiny, much longer than broad
     8 Leaves palmately compound; leaflets usually more than 4 .... **Lupinus**
     8 Leaves not palmately compound
      9 Fruits definitely constricted between the seeds, breaking into 1-seeded segments .... **Hedysarum**
      9 Fruits not constricted between seeds, dehiscing lengthwise
       10 Flowers in small heads or umbels .. **Lotus**
       10 Flowers in spikes or racemes
        11 Keel of corolla abruptly narrowed to a beak-like point; leaves chiefly basal .... **Oxytropis**

11 Keel of the corolla not beaked; plants mostly with leafy stems .... **Astragalus**

ASTRAGALUS Locoweed; Milk Vetch

Annual or perennial herbs. Lvs alt, mostly odd-pinnately compound. Fls borne mostly in axillary racemes, corolla white or yellowish to reddish or purple, the keel usually blunt, not pointed or beaked, stamens 10 diadelphous; fruit a flattened or inflated legume.

1 Plants forming a cushion on the ground, not over 5 dm tall; leaflets less than 1 mm wide, spine tipped . **A. kentrophyta**
1 Plants not forming a cushion; leaflets more than 1 mm wide, not spine tipped
  2 Stems bearing flowers originating at the ground level, unbranched and lacking leaves .... **A. purshii**
  2 Stems bearing leaves and flowers
    3 Pubescence of the leaves dolabriform (see drawing p. 115)
      4 Stipules on the lower nodes united into a bidentate sheath
        5 Fruits erect, completely bilocular .... **A. canadensis**
        5 Fruits oriented in various ways, always unilocular .... **A. miser**
      4 Stipules all free .... **A. terminalis**
    3 Pubescence basifixed
      6 Flowers large, the calyx-tube 5-10 mm, the banner 13.5-24 mm long .... **A. agrestis**
      6 Flowers smaller, the calyx-tube 1.5-4 mm, the banner 4.5-13 mm long
        7 Stems arising from a subterranean root-crown or rhizome-like branches .... **A. alpinus**
        7 Stems arising together from a determinate superficial root-crown or caudex
          8 Lower stipules large, obtuse, several nerved, often only obscurely connate .... **A. aboriginum**
          8 Lower stipules small, definitely united forming a sheath
            9 Fruits stipitate; (has the odor of selenium) .... **A. bisulcatus**
            9 Fruits sessile or nearly so (no selenium odor) .... **A. miser**

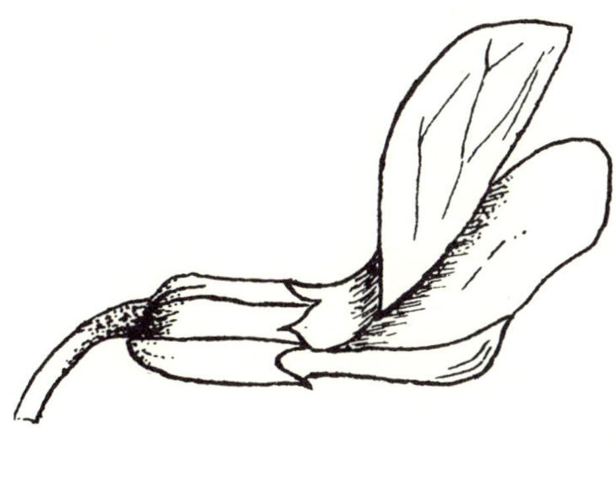

Papilionaceous flower

Exploded view of flower

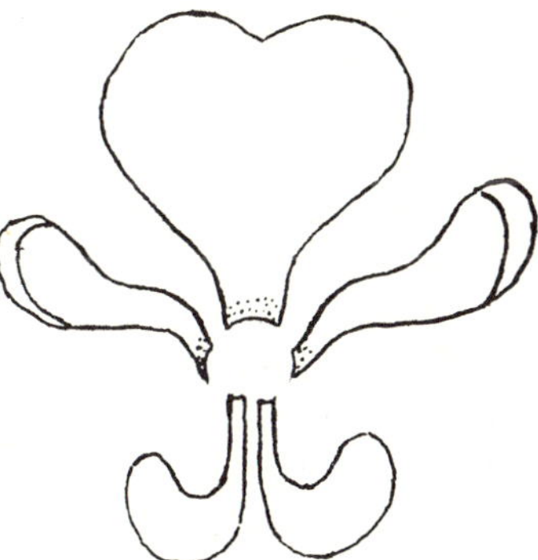

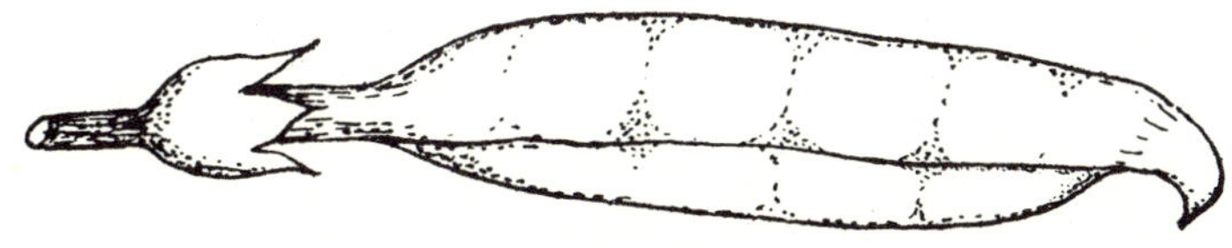

Fruit

Dolabriform hairs on leaf (X section) 20 X

Fabaceae

1. A. **aboriginum** Rich. Never collected within the park, but reported in adjacent areas.
2. A. **agrestis** Dougl. Frequent in loamy soils in meadows near the Oxbow Bend of the Snake River.
3. A. **alpinus** L. var. **alpinus.** Common along the Snake River and in some of the major canyons.
4. A. **bisulcatus** (Hook.) Gray var. **bisulcatus.** Frequent east park boundary; poisonous to grazing animals because of the selenium that it absorbs from the soil.
5. A. **candensis** L. var. **brevidens** (Gand.) Barneby. Sandy river bottom, near Bar BC Ranch.
6. A. **kentrophyta** Gray var. **implexus** (Canby) Barneby. Frequent on rocky slopes near and above timberline.
7. A. **miser** Dougl. var **hylophilus** (Rydb.) Barneby. Frequent along the Snake River.
8. A. **purshii** Dougl. var. **purshii.** Frequent in meadows near Oxbow Bend of the Snake River.
9. A. **terminalis** S. Wats. Common along the dry terraces above the Snake River.

CARAGANA Siberian Pea Tree

Small to large shrub. Lvs even-pinnate. Fls bright yellow, solitary or in fascicles; stamens monadelphous; legumes linear, usually pointed.

†1. **C. arborescens** Lam. Remaining from cultivation east of the Jackson Lake Dam.

GLYCYRRHIZA Licorice

Perennial herbs with glandular dotted foliage. Fls whitish to yellowish, in axillary racemes; stamens diadelphous; fruits covered with uncinate prickles.

1. **G. lepidota** (Nutt.) Pursh. Common along the Snake River bottoms especially below Moose.

HEDYSARUM Sweetvetch

Perennial herbs. Lvs odd pinnate. Fls borne in bracteate racemes, pink-purple, wings shorter than the keel; stamens diadelphous; fruit a loment (definitely constricted between seeds).

1 Upper calyx lobes broader and definitely shorter than the lower 3; petals of the wings with a slender basal lobe nearly as long as the claw; loments sometimes with winged-margins .... **H. occidentale**
1 Upper calyx lobes slender, about equal to the lower ones, basal lobe of wing petals shorter than the claw; loments wingless .... **H. boreale**

1. **H. boreale** Nutt. var. **boreale.** Locally frequent in gravel south of Spread Creek Bridge on Hwy. 89.
1. **H. occidentale** Greene. Frequent in upper So. Cascade Canyon and slopes of Symmetry Spire.

LOTUS Bird's Foot Trefoil

Annual or perennial herbs. Lvs pinnately compound, usually with 3 leaflets. Fls in 1-to many flowered umbels, white or yellow but often tinged with reddish-purple; stamens diadelphous; legumes 1-to many seeded.

†1. **L. corniculatus** L. Locally frequent in disturbed sites at the Jackson Lake Lodge.

LUPINUS Lupine

Annual or perennial herbs sometimes becoming woody toward the base. Lvs alt, palmately compound, 5-17 leaflets. Fls borne in racemes, corolla white, blue or violet (in ours); calyx usually bilabiate; banner with reflexed sides, glabrous to pubescent on the back; stamens 10, monadelphous; legumes flattened, hairy. A critical genus with specific boundaries that are obscured by rather free interbreeding.

1 Uppermost petal (banner) hairy on outer surface .. **L. sericeus**
1 Uppermost petal glabrous or nearly so
  2 Plants nearly acaulescent, stem rarely over 1 dm long and usually shorter than the longer petioles ...... **L. lepidus**
  2 Plants caulescent; stem usually over 2 dm long, exceeding the lower petioles
    3 Uppermost petal slightly or moderately reflexed from wings and keel; stem not hollow ..... **L. argenteus**
    3 Uppermost petal well reflexed from wings and keel; stem hollow at least at the base .... **L. polyphyllus**

1. **L. argenteus** Pursh ssp. **parviflorus** (Nutt.) Phillips. Common in the shade of lodgepole, throughout the valley. This species varies in height, pubescence and flower size depending on ecological conditions.
2. **L. lepidus** Dougl. var. **utahensis** (Wats.) Hitchc. Infrequent, open sites, Hedrick's Pond, Jackson Hole Hwy. and Colter Bay Amphitheater.
3. **L. polyphyllus** Lindl. Stream banks and moist forests of the valley. The most lush and rank growing species.
4. **L. sericeus** Pursh var. **sericeus.** Abundant in the sagebrush throughout the valley.

MEDICAGO Alfalfa; Medick

Annual or perennial herbs. Lvs trifoliate, leaflets toothed at least towards the apex. Fls 2-many in spikelike racemes or heads on short peduncles, yellow or purplish-blue; stamens 10, diadelphous; fruit an indehiscent legume, curved or spirally coiled at maturity.

1 Plants annual; flowers yellow, 2-3 mm long ...... **M. lupulina**
1 Plants perennial; flowers purplish-blue, 6-10 mm long .... **M. sativa**

†1. **M. lupulina** L. Black Medick. Common in waste places throughout the valley.
†2. **M. sativa** L. Alfalfa. An escape from cultivation, throughout the valley.

MELILOTUS Sweet Clover

Sweetish-odored herbs, annual or biennial. Lvs trifoliate, leaflets serrulate. Fls small, papilionaceous, borne in slender racemes, white or yellow; stamens 10, diadelphous; fruit an indehiscent 1-seeded legume.

1 Corolla white .... **M. alba**
1 Corolla yellow .... **M. officinalis**

†1. **M. alba** Desr. White Sweet Clover. Common along the highways.
†2. **M. officinalis** (L.) Lam. Yellow Sweet Clover. Common along the highways.

ONOBRYCHIS

Perennial herbs. Lvs odd-pinnate. Fls papilionaceous in spikelike racemes, pink to lavender (in ours), wing petal much smaller than keel and bannr; stamens 10, monadelphous but upper filaments not totally connate; fruit pod indehiscent, 1-2 seeded, usually more or less prickly.

†**1. O. viciifolia** Scop. European weed, first seen at Teton Science School in 1974.

OXYTROPIS Stemless Loco

Caespitose perennial herbs, closely related to *Astragalus*, but with the keel of the corolla prolonged into a narrow beak; leaves odd-pinnate.

1 Stipules attached to the base of the petiole for only 1-3 mm; legumes pendulous .... **O. deflexa**
1 Stipules attached to the base of the petiole for half their length or more; legumes spreading or erect
 2 Corollas purple or reddish-purple .... **O. lagopus**
 2 Corollas white to yellowish, sometimes the keel is tinged with purple .... **O. campestris**

1. O. campestris (L.) DC. Circumboreal; limestone, head of Death Canyon.
2. O. deflexa (Pall.) DC.
 Flowers bluish-purple; 7-10 flowers in racemes .... var. **foliosa** (Hook.) Barneby
 Flowers pale blue; 10-40 flowered .... var. **sericea** T. & G.
 Var. **foliosa.** Frequent below summit of Table Mt.
 Var. **sericea.** Frequent along the Snake River.
3. O. lagopus Nutt. Infrequent, hill north of Kelly Warm Spring.

TRIFOLIUM Clover

Annual or perennial herbs. Lvs mostly palmately compound; leaflets usually 3, often toothed. Fls borne in umbels, heads or racemes, white, yellow, or pink to red, corolla papilionaceous, persistent after withering; stamens 10, diadelphous; legume indehiscent, 1-to several seeded.

1 Calyx strongly pubescent to villous .... **T. pratense**
1 Calyx glabrous or only with scattered hairs
  2 Flowers 5-9 mm long; heads axillary
    3 Corolla white to slightly pink; plants stoloniferous .... **T. repens**
    3 Corolla usually pink to reddish; plants usually not stoloniferous .... **T. hybridum**
  2 Flowers at least 10 mm long; heads often terminal and solitary .... **T. longipes**

†**1. T. hybridum** L. Common in disturbed sites throughout the valley. Introduced from Europe.

2. T. longipes Nutt. var. **reflexum** A. Nels. Frequent in moist meadows east of Jackson Lake.

†**3. T. pratense** L. Common in disturbed sites throughout the valley. European species.

†**4. T. repens** L. Frequent in disturbed sites throughout the valley.

VICIA Vetch

Annual or perennial herbs with weak trailing to climbing stems. Lvs even pinnate ending in simple to branched tendrils. Fls in axillary racemes or on 1-flowered axillary peduncles; calyx unequally 5-toothed; corolla bluish-purple, wings adherent to the keel; stamens 10, usually diadelphous; legumes compressed.

1. V. americana Muhl. This is an extremely variable species, frequent along the Snake River.

FUMARIACEAE Fumitory Family

Herbs. Lvs alt/basal, dissected, no stip. Infl various. Fls bisex, irreg, ov superior. Ca 2, Co 2 + 2, S 3 + 3, P 2*. Capsule.

1 Outer petals different from each other, 1 petal being spurred or saccate; plants with leafy stem .... **Corydalis**
1 Outer petals identical, both spurred or saccate at the base; plants with leafless peduncles .... **Dicentra**

CORYDALIS Corydalis

Annual, biennial or perennial herbs, often with hollow stems. Lvs several times compound. Fls in racemes or panicles, yellow (in ours); sepals 2, membranous; petals 4, the outer pair dissimi-

Dicentra uniflora Steershead Fumariaceae

lar, inner petals connate; stamens 6, in 2 groups; capsule linear to ovoid.

1. **C. aurea** Willd. Locally frequent in moist to dry soil along the Snake River.

DICENTRA Bleedingheart

Perennial, scapose herbs, with fascicled fleshy roots. Lvs 3 times compound. Fls solitary (in ours), white to pinkish; sepals 2; both outer petals spurred or saccate at the base; stamens in 2 groups; elongated capsule; seeds black and shiny.

1. **D. uniflora** Kell. Steershead. Infrequent on well-drained soils in areas such as Park Hdqts., Jenny Lake and Colter Bay.

GENTIANACEAE Gentian Family

Herbs. Lvs opp, entire, no stip. Infl cymose/fls solit. Fls bisex, reg, ov superior. Ca 4-5, Co 4-5, S 4-5, P 2*. Capsule.

1 Corolla campanulate, salverform to tubular, lobes free of fringed glands . . . . **Gentiana**
1 Corolla rotate, with conspicuous fringed glands on the upper surface
  2 Flowers 4-merous . . . . **Frasera**
  2 Flowers 5-merous . . . . **Swertia**

FRASERA Green Gentian

Perennial herbs. Lvs opp, or whorled, entire, prominently nerved. Fls in an elongated mixed panicle; sepals 4, nearly distinct; corolla rotate, 4-lobed, white or yellowish green, spotted with purple, fringed glands on lobes, stamens 4; capsule septicidal.

1. **F. speciosa** Dougl. Locally frequent to common in meadows of the valley and extending into the canyons and mountain meadows.

GENTIANA Gentian

Annual, biennial or perennial herbs. Lvs opp, petiolate to sessile. Fls solitary or in cymose clusters; calyx 4-5 lobed; corolla usually blue, violet or purple, 4-5 lobed, usually with folds running down from the sinuses; stamens epipetalous; capsule 1-loculed.

1 Plants annual; corollas mostly less than 2 cm long, if longer, then the perianth is 4-merous
  2 Petals deep blue, usually over 2 cm long, fringed; flowers 4-merous; stigma erose .... **G. detonsa**
  2 Petals usually violet to purplish blue, less than 2 cm; flowers sometimes 5-merous; stigma not erose .. **G. amarella**
1 Plants perennial; corollas well over 2.5 cm long and usually 5-merous
  3 Flowers usually solitary; plants glabrous; leaves rarely twice as long as broad .... **G. calycosa**
  3 Flowers usually several; plants usually pubescent; leaves usually twice as long as broad .... **G. affinis**

1. **G. affinis** Griseb. Frequent along the Snake River.
2. **G. amarella** L. Frequent in streambed with mosses on north side of Signal Mt., also south of Colter Bay.
3. **G. calycosa** Griseb. Frequent in meadows and on streambanks in the major canyons above 8,000 ft.
4. **G. detonsa** Rottb. var. **unicaulis** (A. Nels.) C. L. Hitchcock. Frequent along small streams in willow areas north of Jackson Lake Lodge. [*G. thermalis* Kuntze.].

SWERTIA

Annual or perennial herbs. Lvs opp, entire and sessile. Fls paniculate; perianth usually 5-merous; corolla bluish-purple to white or greenish, each segment with a basal pair of glands; stamens 5; capsule with one locule.

1. **S. perennis** L. Felwort. Frequent along small streams in such canyons as Death and Webb Canyons.

GERANIACEAE Geranium Family

Herbs. Lvs alt/opp, simple/compound, usually stip. Infl various. Fls bisex, reg, ov superior. Ca 5, Co 5, S 5-10, P 5*. Capsule/schizocarp.

1 Leaves pinnately compound and dissected; anther bearing stamens 5 .... **Erodium**
1 Leaves palmately lobed or dissected; anther bearing stamens 10 .... **Geranium**

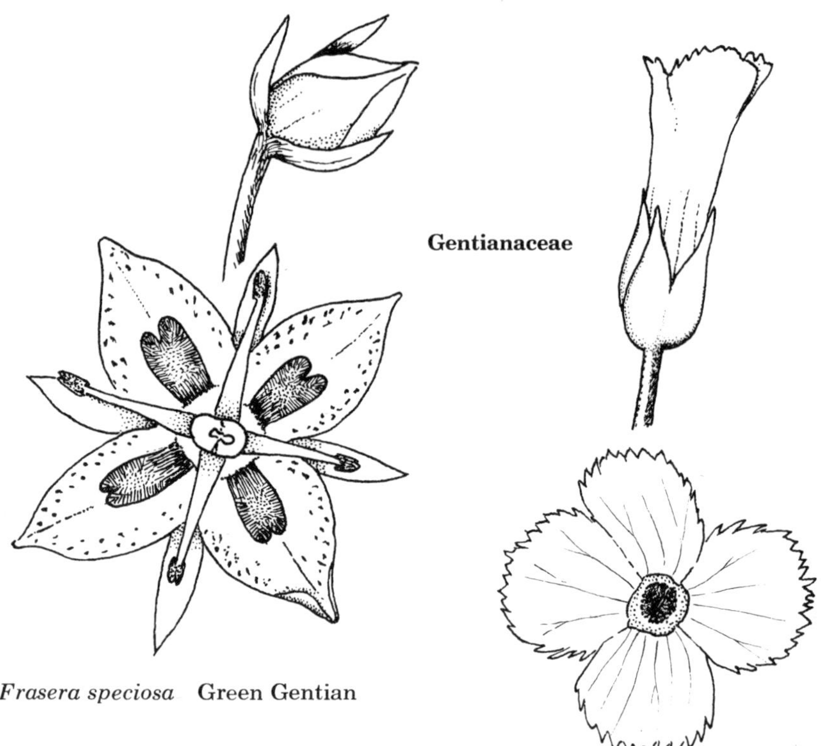

Gentianaceae

Frasera speciosa Green Gentian

Gentiana detonsa Fringed Gentian

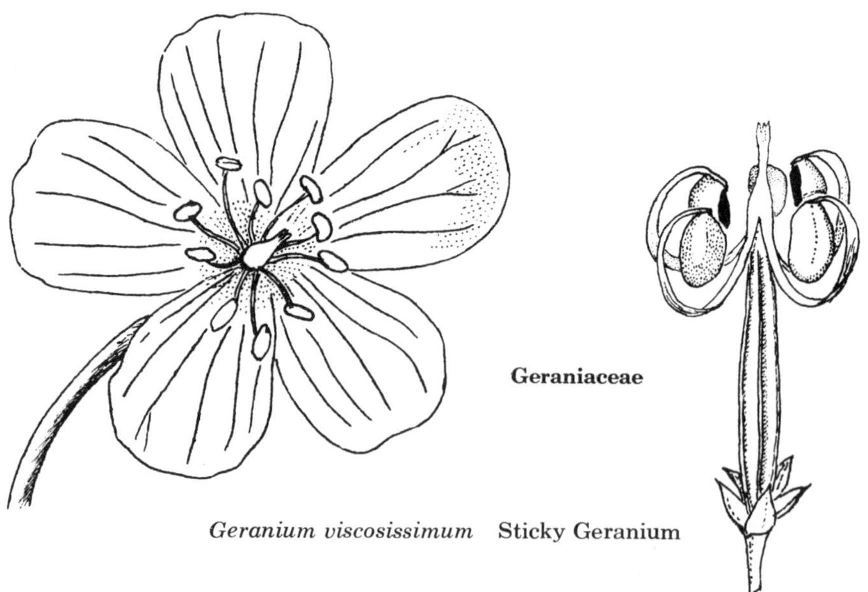

Geraniaceae

Geranium viscosissimum Sticky Geranium

ERODIUM Filaree; Crane's Bill

Annual herbs. Lvs opp, pinnately compound, stipulate. Fls in umbels on axillary peduncles; sepals 5, persistent; petals 5, pink; filaments 10, the 5 shorter ones lacking anthers; styles twisting spirally at maturity.

†1. **E. cicutarium** (L.) L'Her. Infrequent, Eurasian species in disturbed sites.

GERANIUM Geranium

Annual, biennial or perennial herbs. Lvs palmately lobed, stipulate. Fls cymose; 5-merous, sepals persistent; corolla pink or white; stamens mostly 10; carpels 5, at maturity splitting upward.

1 Corollas white with purple veins; plants of wet habitats .... **G. richardsonii**
1 Corollas rose-purple to pink; plants of drier habitats .. .. **G. viscosissimum**

1. **G. richardsonii** Fisch. & Trautv. Frequent on stream banks or moist areas throughout the valley.
2. **G. viscosissimum** F. & M. var. **viscosissimum.** Common along the Snake River and open xeric sites of the moraines.

GROSSULARIACEAE

Shrubs (in ours). Lvs alt, simple, usually no stip. Fls usually bisex, reg, ov inferior (in ours). Ca 4-6, Co 4-6, S 4-6, P 2-6*. Berry.

RIBES Currants and Gooseberries

Erect to spreading shrubs with armed or unarmed stems. Lvs alt, palmately veined, 3-5 lobed and variously toothed. Fls 2 to many in bracteate racemes. Petals 4-6, petals greenish white, white, yellow, or pink to red. Hypanthium tubular to saucer-shaped. Stamens usually 5, alternate with the petals. Fruit a many seeded berry, generally crowned with the persistent floral parts. This is a difficult genus, but there is a clear distinction between the armed species, commonly called gooseberries and the unarmed currants.

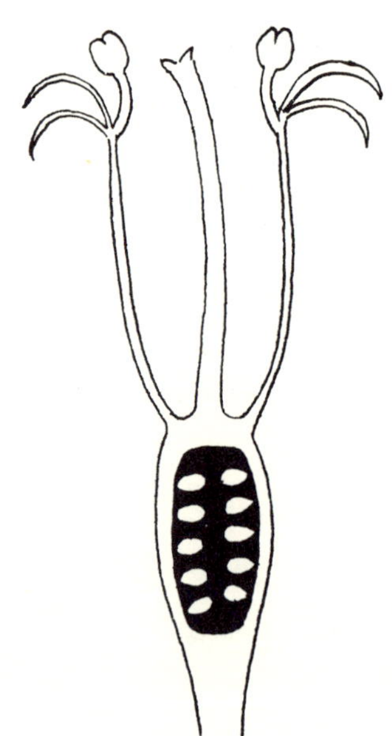

Ribes inerme Common Gooseberry

Grossulariaceae

1 Plants with spines or prickles on the stems and fruit
  2 Hypanthium tubular or campanulate; racemes 2-5 flowered
    3 Stamens at least twice as long as the petals and prominently exserted .... **R. inerme**
    3 Stamens usually about as long as the petals and not prominently exserted .... **R. setosum**
  2 Hypanthium open, saucer-shaped; racemes 3-15 flowered
    4 Leaves bearing glandular hairs, 1-2.5 cm broad; berries reddish .... **R. montigenum**
    4 Leaves glabrous or only weakly pubescent (not glandular), 2-5 cm broad; berries deep purple . **R. lacustre**
1 Plants without prickles or spines
  5 Hypanthium campanulate to cylindric; never saucer-shaped
    6 Hypanthium twice as long as the sepals; berries red .... **R. cereum**
    6 Hypanthium usually equaling the sepals; berries bluish to black .... **R. viscosissimum**
  5 Hypanthium saucer-shaped, broader than long
    7 Petals cream to pinkish; fruit red .... **R. sativum**
    7 Petals white; fruit black .... **R. hudsonianum**

1. **R. cereum** Dougl. North end of Blacktail Butte.
2. **R. hudsonianum** Richards var. **petiolare** (Dougl.) Jancz. Frequent in the major canyons.
3. **R. inerme** Rydb. Frequent in meadows near Oxbow Bend.
4. **R. lacustre** (Pers.) Poir. Frequent in moist woods and stream banks of the canyons and valley.
5. **R. montigenum** McClatchie. Frequent on subalpine talus slopes of the canyons.

†6. **R. sativum** (Reichb.) Syme. Formerly under cultivation near the old town of Moran.

7. **R. setosum** Lindl. Along streams of the valley.
8. **R. viscosissimum** Pursh. Frequent on the north slope of Signal Mt. and the shore of Bradley Lake.

HALORAGACEAE Water-milfoil Family

Mostly aquatic herbs. Lvs alt/whorled, simple/divided, stip/no stip. Fls usually unisex, reg, ov inferior. Ca 0-2-4, Co 0-2-4, S 1-8, P 1-4*. Nut/drupe-like.

1 Flowers mostly unisexual; stamens more than one; submerged leaves finely dissected .... **Myriophyllum**
1 Flowers mostly perfect; stamen one; leaves all entire . **Hippuris**

HIPPURIS Mare's-tail

Aquatic or amphibious herbs with rhizomes. Lvs whorled, linear, entire. Fls single in axil of lvs; perianth apparently lacking; stamen 1; style slender.

1. H. vulgaris L. Circumboreal. Frequent in shallow ponds north of Jackson Lake Dam.

MYRIOPHYLLUM Water-milfoil

Aquatic herbs. Lvs pinnately dissected on flaccid stems. Fls 1 per axil, subtended by 2 tiny bracts; calyx 4 lobed; petals 4 or lacking; stamens 4-8.

1. M. spicatum L. var. **exalbescens** (Fern.) Jeps. Frequent in quiet water; Swan Lake.

HYDROPHYLLACEAE Waterleaf Family

Usually herbs. Lvs alt/basal rarely opp, entire/divided, no stip. Infl helicoid cymes/fls solitary. Fls bisexual, reg, ov superior. Ca 5*, Co 5*, S 5, P 2*. Capsule.

1 Flowers more or less in helicoid cymes; placentae evidently intruded and partition-like .... **Phacelia**
1 Flowers in head-like cymes or solitary; placentae enlarged but not partition-like
  2 Plants perennial; flowers in head-like clusters on peduncles; stamens exserted .... **Hydrophyllum**
  2 Plants annual; flowers solitary (in ours); stamens included .... **Nemophila**

HYDROPHYLLUM Waterleaf

Perennial herbs. Lvs alt and basal, pinnately divided. Fls borne in compact, head-like clusters; calyx campanulate, divided into 5 divisions; corolla campanulate, white to purple; stamens 5, exserted; style 2-cleft; capsule with one locule.

1. H. capitatum Dougl. Frequent on moraines throughout the valley.

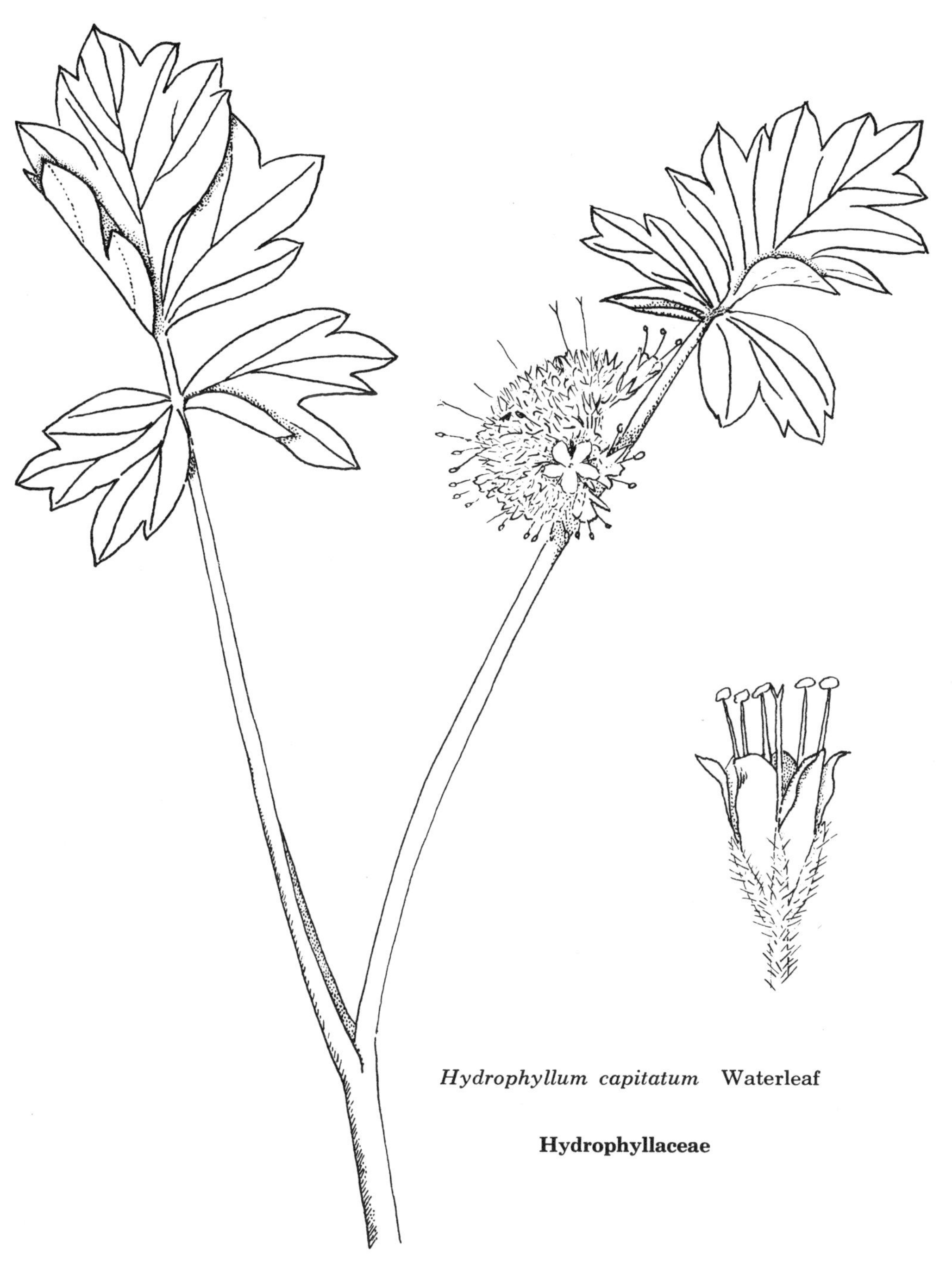

Hydrophyllum capitatum Waterleaf

Hydrophyllaceae

NEMOPHILA Baby-blue-eyes

Delicate taprooted annual herbs. Lvs opp or alt, lobed to pinnatifid. Fls usually solitary; calyx divided to near the base into 5 lobes usually with an appendage in each sinus; corolla campanulate to rotate, white, blue or purple, 5 lobes; stamens 5; capsule 1-loculed, dehiscent by 2 valves.

1. **N. breviflora** Gray. Locally common along the bank of the Snake River near Menor's Ferry.

PHACELIA Scorpion Weed

Annual to perennial taprooted herbs. Lvs mostly alt, entire to pinnately dissected. Fls borne in helicoid cymes; calyx 5-lobed; stamens 5, often long exserted; style 2-cleft; capsule loculicidal.

1 Leaves all entire or some with 2-4 basal lobes
  2 Plants biennial or short-lived perennial from a taproot, usually with a single erect stem, usually over 5 dm tall, sometimes surrounded by smaller stems .... **P. heterophylla**
  2 Plants perennial from a taproot bearing a branched caudex with several stems usually less than 5 dm tall .... **P. hastata**
1 Leaves coarsely toothed or pinnately lobed
  3 Plants annual or biennial; usually with single stems or the central stem surrounded by smaller stems; corolla glabrous within; filaments scarcely exserted ... . **P. franklinii**
  3 Plants perennial, usually several-stemmed; corolla hairy within; filaments long-exserted .... **P. sericea**

1. **P. franklinii** (R. Br.) Gray. Frequent in gravelly soil along the Snake River.
2. **P. hastata** Dougl. var. **alpina** (Rydb.) Cronq. [*P. leucophylla* Torr.]. In talus of subalpine areas.
3. **P. heterophylla** Pursh. Common in disturbed sites of the valley.
4. **P. sericea** (Grah.) Gray. Frequent in rocky places in the major canyons of the range.

HYPERICACEAE St. John's Wort Family

Herbs. Lvs opp, simple, punctate with glandular dots. Infl cymose. Fls bisex, reg, ov superior. Ca 5, Co 5, S ∞, P 3-5*. Capsule.

HYPERICUM St. John's Wort

Perennial herbs, glabrous. Lvs opp, sessile, mostly blackish dotted along the margins. Fls cymose; sepals and petals 5, yellow; pistil 3 (in ours); capsule septicidal.

1 Sepals triangular to ovate-lanceolate, apex rounded to acute; seeds yellowish; native species .... **H. formosum**
1 Sepals linear-lanceolate, apex mostly acute; seeds brownish; introduced alien .... **H. perforatum**

1. H. formosum H. B. K. var. **nortoniae** (Jones) Hitchc. Frequent in the major canyons above 8,000 ft.
†**2. H. perforatum** L. Disturbed sites along Teton Park Road.

LAMIACEAE (LABIATAE) Mint Family

Herbs/shrubs. Lvs opp, aromatic, simple. Stem square. Infl often verticillate. Fls bisex, irreg, ov superior. Ca 5*, Co 5*, S 4, P 2*. Fruit of 4 nutlets.

1 Calyx with a sac-like protuberance on the upper side; lips of calyx entire .... **Scutellaria**
1 Calyx lacking a protuberance; at least one of calyx lips toothed
  2 Corolla scarcely bilabiate, usually only 4-lobed .... **Mentha**
  2 Corolla more or less strongly bilabiate
    3 Stamens obviously visible without dissection of the flower; upper lips of the corolla not forming a hood (galea) .... **Agastache**
    3 Stamens scarcely or not at all exserted; upper lip of corolla forming a hood (galea)
      4 Calyx 5-10 nerved; lower stamens longer than the upper .... **Prunella**
      4 Calyx 15 nerved; upper stamens longer than the lower .... **Dracocephalum**

AGASTACHE Giant Hyssop

Perennial herbs. Lvs opp, petioled, toothed. Fls in bracted, spike-like inflorescences; calyx tubular-campanulate, 5 lobes; corolla pink to violet or white, bilabiate; stamens 4, usually exserted.

1. **A. urticifolia** (Benth.) Ktze. Common on open slopes of moraines of the valley.

DRACOCEPHALUM Dragonhead

Annual or perennial herbs. Lvs opp, toothed or cleft. Fls borne in whorls in the axils of small or leafy bracts; calyx tubular, 5 toothed; corolla blue or purple to sometimes white, bilabiate; stamens 4, rising under the upper lip of the corolla; nutlets ovoid, 3-sided, smooth. (*Moldavica*)

1. **D. parviflorum** Nutt. [*M. Parviflora* (Nutt.) Britt.]. Infrequent along the banks of the Snake River and disturbed sites at Coter Bay.

MENTHA Mint

Aromatic, perennial herbs with rhizomes. Lvs opp, mostly petioled and toothed. Fls in whorls, subtended by leaves or bracts, often forming spike-like inflorescences; calyx campanulate, 5-lobed; corolla violet, funnelform or campanulate, nearly regular; nutlets smooth.

1. **M. arvensis** L. var. **glabrata** (Benth.) Fern. Infrequent in moist places, Snake River and small ponds north of Colter Bay.

PRUNELLA Self-heal

Perennial herbs 5-30 cm tall. Lvs opp, entire to pinnatifid. Fls in whorls forming a terminal spike, with bracts; calyx bilabiate; corolla blue or purple to white, bilabiate; stamens 4, ascending under the upper lip of the corolla.

1. **P. vulgaris** L. Frequent in moist places beside streams and lakes of the valley.

SCUTELLARIA Skullcap

Perennial herbs (in ours). Lvs opp, entire to crenate-dentate. Fls solitary in the axils of slightly reduced leaves; calyx bilabiate, crest on the upper side; corolla blue marked with white (in ours), bilabiate; stamens 4, the lower pair longer; nutlets covered with small projections.

1. **S. galericulata** L. Frequent in wet meadows north of Jackson Lake Dam, circumboreal.

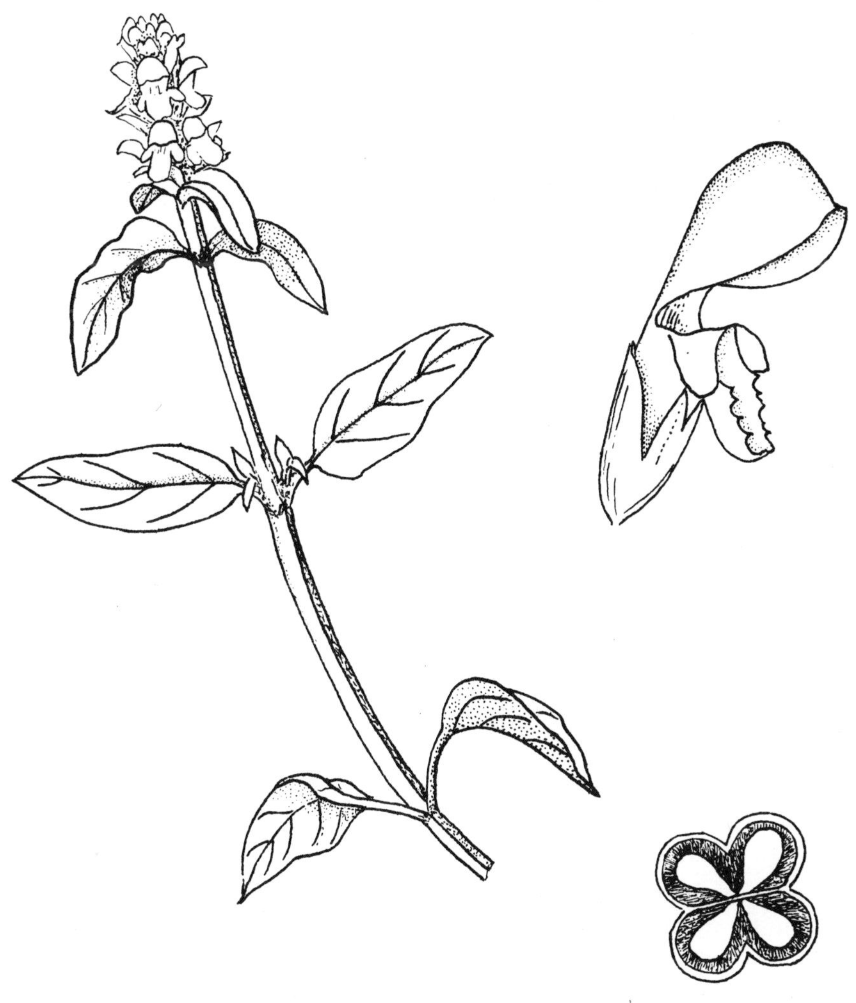

Prunella vulgaris Self-heal **Lamiaceae**

LENTIBULARIACEAE Bladderwort Family

Aquatic or semiaquatic herbs. Lvs alt/basal, divided and usually bearing small inflated bladders. Infl scapose, raceme/fls solitary. Fls bisex, irreg, ov superior. Ca 2-5*, Co 5*, S 2, P 2*. Capsule.

UTRICULARIA Bladderwort

Submersed aquatic herbs. Lvs alt, finely dissected, submersed lvs provided with buoyant bladders which serve as traps form small crustaceans. Fls solitary or racemes; calyx 2-lobed; corolla yellow, bilabiate, with a spur at the base; capsule with numerous seeds.

1 Lower lip of corolla mostly 10-20 mm long; leaves mostly 2-parted at the base; spur of corolla well developed, directed forward .... **U. vulgaris**
1 Lower lip of corolla mostly 4-12 mm long; leaves mostly 3-parted at the base; spur of corolla poorly developed, about half as long as the lower lip .... **U. minor**

1. **U. minor** L. Rare in standing or slowly moving water south side of Leigh Lake.
2. **U. vulgaris** L. Rare in old beaver pond ¼ mile north of Jackson Lake Lodge.

LIMNANTHACEAE Meadow-foam

Herbs of wet places. Lvs alt, divided, no stip. Fls solitary, bisex, reg, ov superior. Ca 3-5, Co 3-5, S 6-10, P 3-5*. Achene.

FLOERKEA False-mermaid

Juicy herbs of moist areas. Fls small; petals less than 2 mm; perianth usually 3-merous; stamens 3-6; fruit covered with wart-like projections.

1. **F. proserpinacoides** Willd. Infrequent in wet places, especially under shrubs, in the valley.

LINACEAE Flax Family

Herbs. Lvs alt/opp, entire, stip/no stip. Infl cymose. Fls bisex, reg, ov superior. Ca 5(*), Co 5 soon falling, S 5, P 5. Capsule.

LINUM Flax

Characteristics of the family. Petals blue (in ours); capsule septicidal; seed sometimes mucilaginous when wet.

1. **L. perenne** L. var. **lewisii** (Pursh) Eat. & Wright. Locally frequent in relatively dry soil in the valley. Also infrequent in upper area of South Cascade Canyon.

LOASACEAE Loasa Family

Herbs often with stiff hairs and whitish stems. Lvs opp/alt, simple/divided, no stip. Fls axillary, bisex, reg, ov inferior. Ca 4-5, Co 5-10, S ∞, P 3-7*. Capsule.

MENTZELIA Blazing-star

Annual to perennial, coarse herbs. Lvs alt, rough to the touch, with minute barbs. Fls small to showy, yellow to orange.

1 Plants annual; petals less than 1.5 cm long; seeds not flattened .... **M. dispersa**
1 Plants biennial to perennial; petals 2-7 cm; seeds flattened .... **M. laevicaulis**

1. **M. dispersa** Wats. Locally frequent on dry slopes throughout the valley.
2. **M. laevicaulis** (Dougl.) T. & G. var. **laevicaulis.** Infrequent on gravelly soil, particularly roadside cuts.

LORANTHACEAE Mistletoe Family

Parasites on branches. Lvs usually opp, simple, no stip, leathery. Fls uni/bisex, reg, ov inferior. Per 2-5, S 2-5, P 3-5*. Berry.

ARCEUTHOBIUM Dwarf Mistletoe

Parasitic, dwarf shrubs, on the branches of conifers. Lvs opp, reduced to connate scales. Fls yellow to greenish; plants dioecious (in ours); more or less fleshy tepals; fruit a mucilaginous berry, the seed expelled explosively.

1 Plants parasitic on *Pseudotsuga menziesii* .... **A. douglasii**

Linum perenne Flax

Linaceae

Arceuthobium americanum

Dwarf Mistletoe ♀ plant

Loranthaceae

1 Plants parasitic on *Pinus* spp.
  2 Plants parasitic on *P. contorta* .... **A. americanum**
  2 Plants parasitic on *P. flexilis* and *P. albicaulis* ....
  .... **A. cyanocarpum**

1. **A. americanum** Nutt. Frequent parasite on Lodgepole Pine throughout the valley.
2. **A. cyanocarpum** Coulter & Nelson. Not seen in G.T.N.P., but reported in areas nearby.
3. **A. douglasii** Engelm. Not seen in G.T.N.P., but reported on Teton Pass Road, 4 miles east of Idaho boundary.

MALVACEAE Mallow Family

Herbs/shrubs often with stellate hairs. Lvs alt, simple/ divided, stip. Infl various. Fls usually bisex, reg, ov superior. Ca 5*, Co 5, S ∞*, P 2-∞*. Capsule.

ILIAMNA Globemallow

Perennial herbs. Lvs alt, 3-9 palmately lobed. Fls racemose, calyx subtended by 3 bracts; petals pink to lavender; carpels dehiscent, hairy with stiff hairs.

1. **I. rivularis** (Dougl.) Greene. var. **diversa** (Nels.) Hitchc. Common along the roads of the valley.

MENYANTHACEAE Buckbean Family

Aquatic/marsh herbs. Lvs alt, entire/trifoliate, petioles sheathing. Infl various. Fls bisex, reg, ov superior. Ca 5*, Co 5, S 5, P 2*. Capsule.

MENYANTHES Buckbean

Rhizomatous, aquatic herbs. Lvs alt, basal, the long petioles with long sheathing stipules, blade trifoliate. Fls borne on a scape, racemose or paniculate; calyx 5-lobed; corolla white or pinkish, 5-6 lobed; stamens 5; indehiscent capsule.

1. **M. trifoliata** L. Locally common in ponds and small lakes, north east portion of the park.

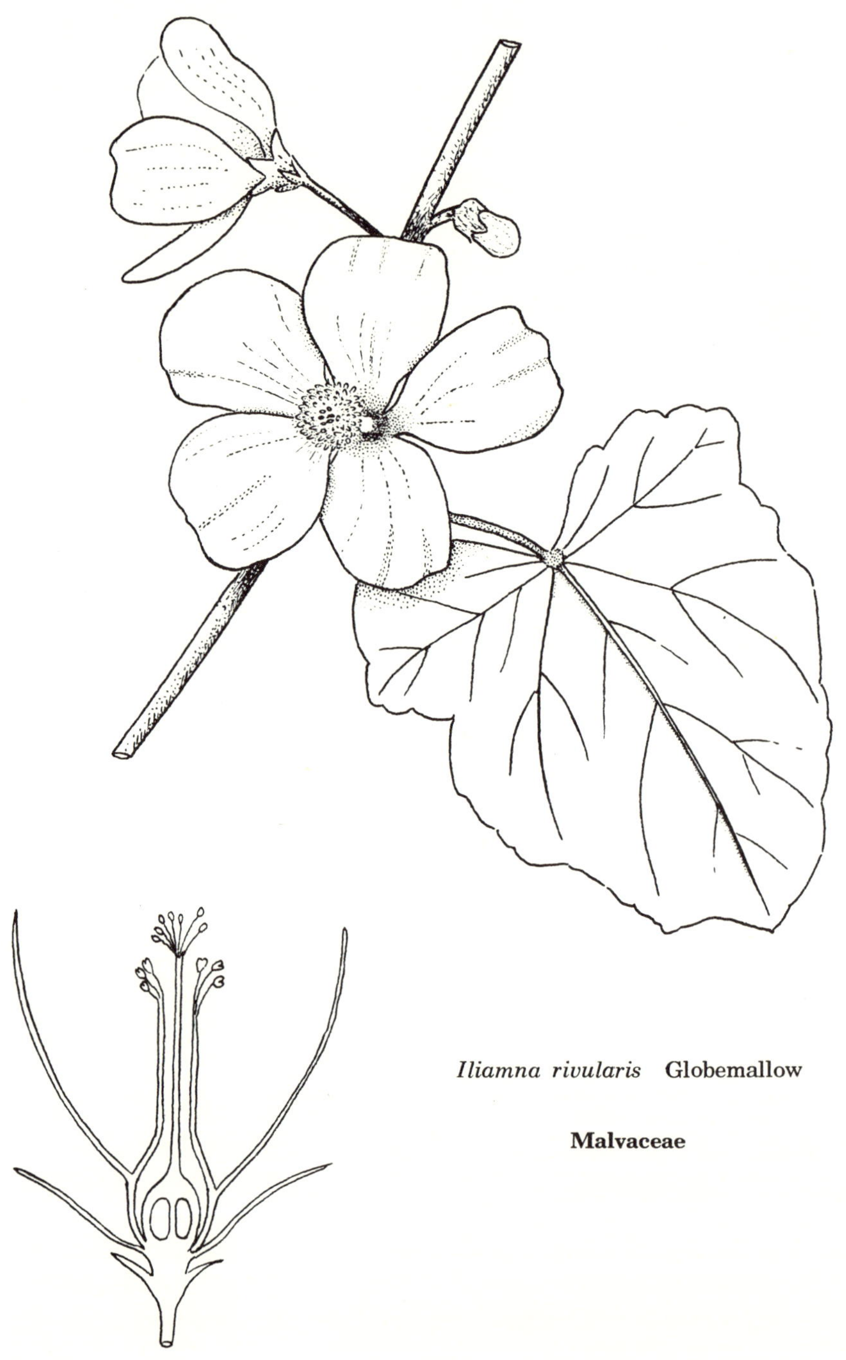

Iliamna rivularis Globemallow

Malvaceae

MORACEAE Mulberry Family

Woody/herbs (twining vines in ours). Lvs alt, simple, entire/serrate/lobed, stip. Infl various. Fls unisex, reg, ov superior. Ca 4-5, Co 0, S 1-4, P 2*. Multiple fruit.

HUMULUS Hop

Characteristics of the family

†**1. H. lupulus** L. Persisting from cultivation at some old ranches of the valley.

NYMPHAEACEAE Waterlily Family

Aquatic, perennial plants. Floating or emergent lvs alt, simple. Fls solitary, bisex, reg, ov superior. Ca 3-∞, Co 3-∞, S ∞, P 1-∞, Fruit various.

NUPHAR Cow Lily; Yellow Water Lily

Aquatic, perennial herbs with fleshy rhizomes. Lvs large, usually floating, cordate. Fls borne on long peduncles; sepals 5-12, yellow or greenish; petals 10-20, smaller than the sepals, often stamen-like; stamens numerous; pistil with several locules, stigma flattened and having a scalloped margin; fruit a berry-like capsule.

1. N. polysepalum Engelm. Common in beaver ponds and shallow lakes of the valley.

ONAGRACEAE Evening Primrose Family

Usually herbs, Lvs alt/opp, simple, stip/no stip. Fls solitary/racemose, bisex, usually reg, ov inferior. Ca 4, Co 4, S 8, P 4*. Capsule/berry/nut.

1 Flower parts in twos; fruits with hooked hairs; leaves broadly ovate, with long petioles .... **Circaea**
1 Flower parts in fours; fruits not bristly; leaves narrower, with short petioles or sessile
  2 Seeds with a tuft of hairs at one end
    3 Flowers scarlet; hypanthium 2-3 cm long and funnelform .... **Zauschneria**
    3 Flowers never scarlet; hypanthium less than 1 cm long or lacking .... **Epilobium**

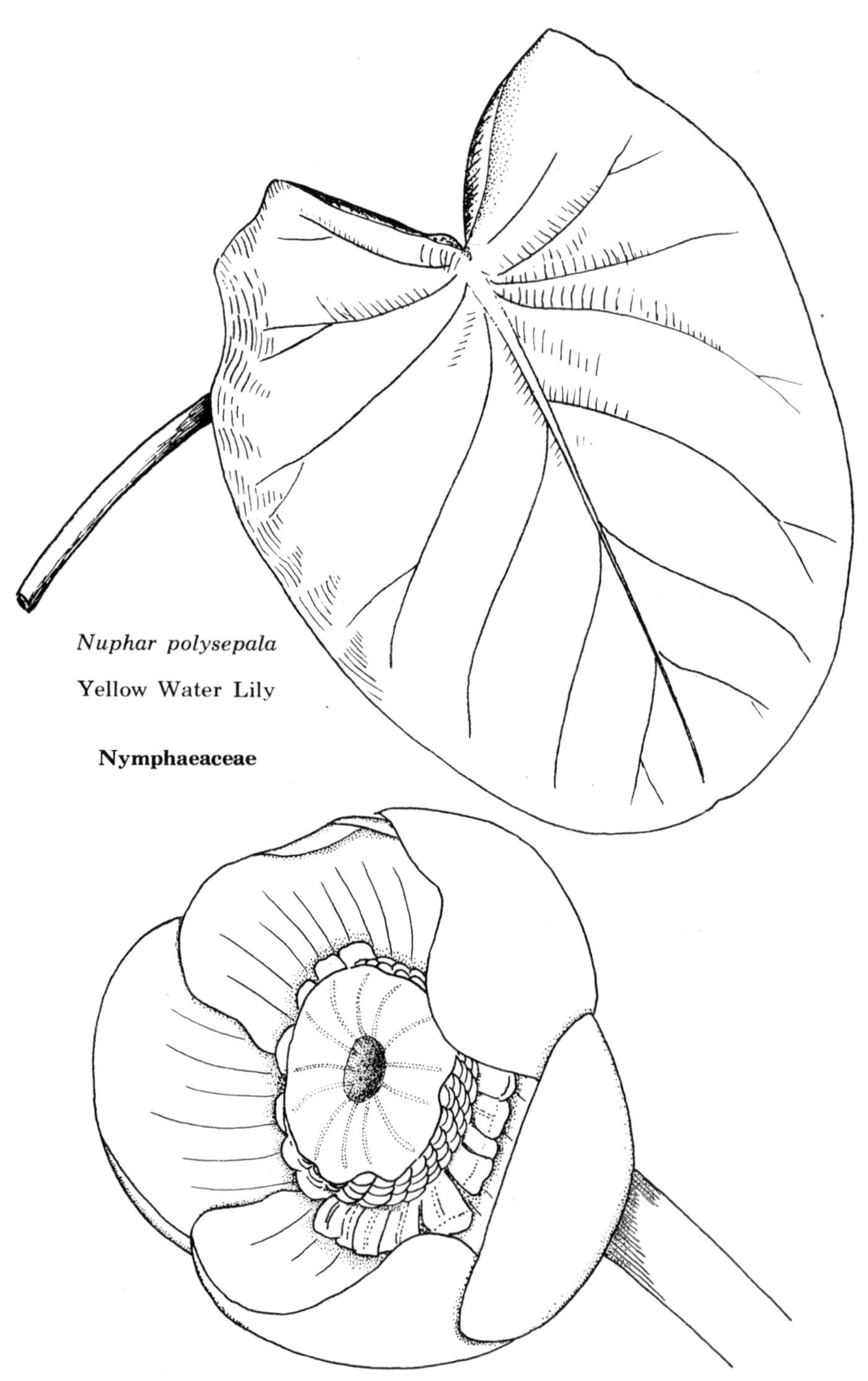

Nuphar polysepala

Yellow Water Lily

Nymphaeaceae

2 Seeds without a tuft of hairs
4 Ovary 2-loculed; plants very delicate, with slender branches; hypanthium not prolonged beyond the ovary .... **Gayophytum**
4 Ovary 4-loculed; plants moderately stout; hypanthium prolonged beyond the ovary as a slender tube
5 Stigma capitate, entire .... **Camissonia**
5 Stigma 4-lobed .... **Oenothera**

CAMISSONIA Evening Primrose

Annual or perennial herbs, from a taproot. Lvs basal or cauline, alternate. Fls lasting less than a day, mostly 4-merous; petals yellow (in ours); anther versatile or basifixed; stigma capitate or hemispherical; fruit a capsule.

1. **C. breviflora** (T. & G.) Raven. Locally frequent in mud, north end of Swan Lake.
2. **C. subacaulis** (Pursh) Raven. Frequent in meadow, usually where dry by late summer.

CIRCAEA Enchanter's Nightshade

Perennial herbs. Lvs opp. Fls small, 2-merous, borne in compound racemes; petals white, notched; stamens alternate with the petals; fruit nutlike 1 or 2 seeded, usually with hooked hairs.

1. **C. alpina** L. Frequent in deep shade, Death Canyon Trail at Phelps Lake level.

EPILOBIUM Willow Herb

Annual or perennial herbs. Lvs opp, sessile or petiolate, entire to dentate. Fls borne in simple or compound racemes; petals white or yellowish to deep rose purple; stamens 8; stigma 4 lobed; fruit a capsule.

1 Flowers large, petals 1-2 cm long, spreading, entire
2 Leaves usually over 8 cm long; plants mostly over 4 dm tall; styles exceeding the stamens .... **E. angustifolium**
2 Leaves less than 8 cm long; plant usually less than 4 dm tall; styles shorter than the stamens .... **E. latifolium**
1 Flowers smaller, the petals usually notched, ascending
3 Petals yellow; plants with a woody base .. **E. suffruticosum**
3 Petals other than yellow; plants not woody at base
4 Plants annual, taprooted, often in dry soil

5 Leaves mostly opposite; plants mostly less than 4 dm tall . . . . **E. minutum**
5 Leaves mostly alternate, except for lowermost; plants usually over 4 dm tall . . . . **E. paniculatum**
4 Plants perennial, usually with rhizomes, often in moist soil
6 Plants forming bulblike offsets . . . . **E. glandulosum**
6 Plants not forming bulblike offsets
7 Stems usually under 3 dm tall, simple or with a few basal branches; rhizomes usually well developed . . . . **E. alpinum**
7 Stems usually over 3 dm tall and branched above the middle; rhizomes reduced or lacking . . . . **E. watsonii**

1. **E. alpinum** L. Frequent in couloirs and canyons above 8,000 ft., often above timberline.
2. **E. angustifolium** L. Fireweed. Abundant along roadside and old fire burns.
3. **E. glandulosum** Lehm. Frequent in aspen groves along the Snake River.
4. **E. latifolium** L. Frequent by streams in Garnet and South Cascade Canyons.
5. **E. minutum** Lindl. Frequent on Phelp's Lake moraine.
6. **E. paniculatum** Nutt. Frequent in sagebrush, Blacktail Butte, and slopes of Signal Mt.
7. **E. suffruticosum** Nutt. Frequent in the gravel of the Snake River bottom.
8. **E. watsonii** Barbery. [*E. adenocaulon* **Hausskn.**]. Common in wet places, Oxbow Bend.

GAYOPHYTUM Ground Smoke

Slender stemmed annuals. Lvs mostly alt, linear to narrowly lanceolate. Fls small; petals white to pinkish; stamens 8, unequal; stigma capitate; fruit a capsule.

1. **G. nuttallii** T. & G. Frequent; dry gravelly soils in the valley.

OENOTHERA Evening Primrose

Annual to perennial, acaulescent to caulescent herbs. Lvs basal or alternate, simple to deeply dissected. Fls borne in spikes, racemes or sessile among basal leaves; petals white or yellow,

Epilobium angustifolium Fireweed

Onagraceae

frequently aging to pink or purple; stamens 8, anthers usually versatile; 4 locules; seed numerous.

1 Plants without conspicuous leafy stems; hypanthium several times as long as the ovary ....**O. caespitosa**
1 Plants with conspicuous leafy stems; hypanthium often shorter than the mature fruit
 2 Petals white, aging to pink, 1.5-3 cm long .... **O. pallida**
 2 Petals yellow, aging to orange or reddish-purple; 2.5-4 cm long .... **O. hookeri**

1. **O. caespitosa** Nutt. Infrequent in roadside cuts and dry hills of the valley.
2. **O. hookeri** T. & G. var. **ornata** (Nels.) Munz. An extremely variable and complex species usually on roadside cuts.
3. **O. pallida** Linkl. Infrequent on west side of Blacktail Butte.

ZAUSCHNERIA Wild Fuchia; Firechalice

Perennial herbs, somewhat woody at the base. Lvs opp or alt. Fls borne in a spike-like raceme.

1. **Z. garrettii** A. Nels. Frequent on rocky ledges in such canyons as Death, Waterfalls and Cascade, up to 7,500 ft.

OROBANCHACEAE Broomrape Family

Parasitic herbs without green foliage. Lvs alt, scalelike. Infl racemose. Fls bisex, irreg, ov superior. Ca 4-5*, Co 5*, S 4, P 2*. Capsule.

OROBANCHE Broomrape; Cancer-root

Parasitic on the roots of other plants. Lvs without chlorophyll and reduced to scales. Fls borne in spikes, panicles or solitary; calyx 4-5 lobed; corolla bilabiate; stamens 4; stigma entire or 2-4 lobed; capsule with numerous seeds.

1 Flowers sessile, or nearly so, provided with a pair of bractlets just beneath the calyx .... **O. ludoviciana**
1 Flowers borne on long pedicels, bractlets absent
 2 Flowers violet, solitary or few, pedicels arising from ground level .... **O. uniflora**
 2 Flowers dull yellow or purplish; pedicels arising from an erect fleshy stem above the ground ...... **O. fasciculata**

1. **O. fasciculata** Nutt. Infrequent as a parasite on big sagebrush throughout the valley.
2. **O. ludoviciana** Nutt. Infrequent in meadows near the Oxbow Bend of the Snake River.
3. **O. uniflora** L. var. **minuta** (Suksd.) Beck. Rare in wet mossy places in the mouth of Cascade Canyon.

OXALIDACEAE Woodsorrel Family

Herbs. Lvs palmately 3-foliate, alt/basal. Infl cymose/umbelliform/fls solitary. Fls bisex, reg, ov superior. Ca 5, Co 5, S 10*, P 5*. Capsule.

OXALIS Woodsorrel

Characteristics of the family

†1. **O. dillenii** Jacq. Rare along Snake River below dam; petals yellow.

PAEONIACEAE Peony Family

Herb/woody. Lvs alt, ternately compound, no stip. Fls solitary, bisex, reg, ov superior. Ca 5, Co 5 or more, S ∞, P 2-5. Follicle.

PAEONIA Peony

Characteristics of the family.

1. **P. brownii** Dougl. Rare at south end of Jenny Lake near boat dock, also in sagebrush at junction of Taggart Creek and Teton Park Road.

PLANTAGINACEAE Plantain Family

Herbs. Lvs alt/opp/basal. Infl usually spicate. Fls uni/bisex, reg, ov superior. Ca 4*, Co 4*, S 4 or 2, P 1-4*. Capsule/nut.

PLANTAGO Plantain

Characteristics of the family.

1 Plants often woolly at the base; leaves narrowly ovate to lanceolate, gradually tapering into petiole ...... **P. lanceolata**

1 Plants not woolly at the base; leaves ovate to broadly ovate, abruptly contracted at the base .... **P. major**

†**1. P. lanceolata** L. Locally common weed of roadsides and disturbed sites.

†**2. P. major** L. Locally common weed of roadsides, lawns and other disturbed sites.

POLEMONIACEAE Phlox Family

Herbs. Lvs alt/opp, entire/ pinnately divided, no stip. Infl cymose to capitate. Fls bisex, usually reg, ov superior. Ca 5*, Co 5*, S 5 epipetalous, P 3*. Capsule.

1 Plants small annual; leaves reduced to basal whorl beneath compact inflorescence and two persistent cotyledons ... . **Gymnosteris**
1 Plants annual to perennial; leaves well developed with clusters at the base or distributed along the stem or both
  2 Calyx tube of uniform texture, not ruptured by the developing capsule
    3 Leaves entire; calyx tube papery at anthesis; annual (in ours) .... **Collomia**
    3 Leaves pinnately compound; calyx tube herbaceous at anthesis; perennial (in ours) .... **Polemonium**
  2 Calyx tube with green ribs and transparent intervals or, if the calyx tube is all green, then the leaves are sessile and palmately dissected into linear segments
    4 Filaments unequally inserted on the corolla tube; leaves entire, mostly opposite
      5 Plants annual, (upper leaves alternate) .. **Microsteris**
      5 Plants perennial .... **Phlox**
    4 Filaments equally or nearly equally inserted on the corolla tube; leaves seldom both opposite and entire
      6 Leaves sessile, palmately dissected into linear segments
        7 Plants annual; when wet, seeds become mucilaginous .... **Linanthus**
        7 Plants perennial; wet seed remain unchanged
          8 Leaves firm and prickly; plants shrubby .. .. **Leptodactylon**

8 Leaves rather soft; plants woody only at the base .... **Linanthastrum**

6 Leaves of various types, but not both sessile and palmately dissected .... **Gilia**

COLLOMIA Collomia

Annual or perennial herbs. Lvs mostly alt and entire (in ours). Fls borne in headlike clusters; calyx tube papery in texture; corolla pink (in ours); stamens 5, unequally inserted on the tube of the corolla; seeds 1-3 per locule, becoming mucilaginous when wet (in ours).

1. **C. linearis** Nutt. Common in the sagebrush flats and moraines throughout the valley.

GILIA Gilia

Taprooted annual to perennial herbs. Lvs alt, entire to pinnatifid. Fls borne in determinate inflorescences; calyx lobes nearly equal, commonly ruptured by the developing capsule; corolla funnelform or salverform, variously colored (usually red in ours); seed 1-many per locule.

1. **G. aggregata** (Pursh) Spreng. var. **aggregata** [*Ipomopsis aggregata* (Pursh) V. Grant.]. Common in gravelly soils, usually associated with big sagebrush, throughout the valley.

GYMNOSTERIS Gymnosteris

Annual herbs with persistent cotyledons. Lvs usually basally connate, forming an involucre beneath the flower cluster. Fls sessile in a terminal cluster; calyx mostly scarious; corolla white, yellow or pink; filament very short, attached at the corolla throat; several seeds in each locule, becoming mucilaginous when wet.

1. **G. parvula** Heller. Frequent in sagebrush on outwash plain between Jenny Lake and Snake River.

LEPTODACTYLON Leptodactylon

Taprooted, shrubby perennials. Lvs alt or opp, sessile and rigid, palmatifid with narrow, spine tipped segments. Fls sessile, solitary in the axils; calyx with awl-shaped teeth; corolla white to pink or salmon, with long slender tube; stamens equally inserted

Gilia aggregata Skyrocket Gilia **Polemoniaceae**

on the corolla tube; seeds several, remaining unchanged when moistened.

1. **L. pungens** (Torr.) Nutt. Prickly Phlox. Infrequent in the sagebrush south of Signal Mt.

LINANTHASTRUM

Perennial herbs with a woody base. Lvs opp, sessile, palmatifid with linear segments. Fls borne in leafy cymose inflorescences; corolla salverform, white or creamy; locule with 2-4 ovules. This genus is intermediate between *Linanthus* and *Leptodactylon.*

1. **L. nuttallii** (Gray) Ewan. Frequent in Death Canyon and southern peaks of the range above 7,500 ft.

LINANTHUS Linanthus

Taprooted, annual herbs. Lvs mostly opp, sessile, palmatifid with linear segments. Fls borne in determinate inflorescences; corolla variously colored, campanulate to salverform; stamens mostly equal; seeds 1-several per locule.

1 Corolla 1.5-2.5 mm long; seeds solitary in each locule .... **L. harknessii**
1 Corolla 2.5-10 mm long; seeds 2-8 in each locule .... **L. septentrionalis**

1. **L. harknessii** (Curran) Greene. Frequent on the Phelp's Lake moraine.
2. **L. septentrionalis** Mason. Infrequent along the Snake River.

MICROSTERIS

Annual herbs. Lvs opp and alt, entire. Fls small, usually in pairs, white to pinkish; corolla salverform, with slender tube and short, spreading lobes; stamens unequally inserted; seeds one per locule, becoming mucilaginous when wet. A single, highly variable species.

1. **M. gracilis** (Hook.) Greene. Frequent in open places throughout the valley.

PHLOX Phlox; Wild Sweet William

Perennial herbs (in ours). Lvs opp, entire, often narrow and needle-like. Fls borne in cymes or solitary; calyx with thin, non green, transparent intervals between the prominent veins; corolla salverform, white to pink, purple or blue; stamens unequally placed in the corolla tube; seeds 1-4 per locule, not affected when moistened. A difficult genus in which the taxa are not sharply delimited.

1 Plants more or less erect with well developed internodes; leaves often more than 3.5 cm long .... **P. longifolia**
1 Plants compact and a cushion-like habit; leaves short and crowded, mostly under 3 cm long
  2 Leaves narrowly linear, 4-10 mm long and about 0.5 mm wide, often with soft woolly hairs .... **P. hoodii**
  2 Leaves usually longer or wider, soft woolly hairs lacking
    3 Styles 2-5 mm long; calyx usually glandular hairy .... **P. pulvinata**
    3 Styles mostly 5-12 mm long; calyx not glandular hairy .... **P. multiflora**

1. **P. hoodii** Rich. Frequent in the sagebrush of the valley.
2. **P. longifolia** Nutt. Frequent in the big sagebrush throughout the valley.
3. **P. multiflora** A. Nels. Frequent, Hwy 89, near north boundary.
4. **P. pulvinata** (Wherry) Cronq. Frequent in rocky places at moderate to high elevations in the mountains.

POLEMONIUM Jacob's Ladder; Sky Pilot

Perennial herbs (in ours). Lvs alt, pinnately compound or deeply pinnatifid; plants often glandular and offensive to smell. Fls borne in cymose inflorescences; calyx somewhat enlarging in fruit; corolla campanulate to funnelform, mostly blue, purple or white; stamens about equally inserted; seed 1-10 per locule.

1 Corolla funnelform, longer than wide, the lobes shorter than the tube; plants strongly glandular, alpine and subalpine .... **P. viscosum**
1 Corolla campanulate, about as wide as long, the lobes as long as the tube; plants weakly glandular; plants of the valley
  2 Stems solitary, 4-10 dm tall; wet places ..... **P. occidentale**
  2 Stems more or less clustered, 2-4 dm tall; moist to dry places .... **P. pulcherrimum**

1. **P. occidentale** Greene. Skunk Flower. Frequent in wet areas, Sawmill Ponds, east of Jackson Lake Lodge, and Natl. Elk Refuge.
2. **P. pulcherrimum** Hook. var. **pulcherrimum.** Infrequent in moist or shaded places of the valley and canyons.
3. **P. viscosum** Nutt. Frequent in rocky places above 9,000 ft.

POLYGONACEAE Buckwheat Family

Herbs/woody. Lvs alt/opp/whorled, stip sheathing or absent. Infl often cymose. Fls usually bisex, reg, ov superior. Per 3-6, S 2-9, P 2-3*. Achene.

1 Plants with conspicuous sheathing stipules
  2 Leaf blades reniform; perianth deeply 4-parted; ovary strongly compressed .... **Oxyria**
  2 Leaf blades various but not reniform; perianth 5-6 parted; ovary not strongly compressed
    3 Perianth segments usually 5 .... **Polygonum**
    3 Perianth segments usually 6 .... **Rumex**
1 Plants without stipules .... **Eriogonum**

ERIOGONUM Wild Buckwheat

Annual to perennial herbs or subshrubs. Lvs alt to whorled, entire, no stipules. Fls solitary or in clusters; perianth white to yellow, or pink, 6 parted; stamens 9; pistil with 3 carpels; achene. The species are difficult to identify requiring a special terminology and some skill in dissection. See Hitchcock & Cronquist, "Fl. of Pac. N. W." pp 79.

1 Plants annual or biennial .... **E. cernuum**
1 Plants perennial
  2 Bracts leafy in texture, 2-several; perianth appearing to form a stipe (see drawing pp. 154)
    3 Involucral bracts generally reflexed or spreading, lobes at least half as long as the tube
      4 External surface of perianth with hairs .... **E. caespitosum**
      4 External surface of perianth glabrous
        5 Flower stem generally has a whorl of bracts at the midlength position; leaves with grayish

hairs on both surfaces, generally linear or oblanceolate .... **E. heracleoides**

5 Flower stem leafless at midlength; leaves usually elliptic .... **E. umbellatum**

3 Involucral bracts erect, lobes usually less than half as long as the tube .... **E. flavum**

2 Bracts usually scale-like and 3-parted; perianth not appearing to form a stipe (see drawing p. 154) .... **E. ovalifolium**

1. **E. caespitosum** Nutt. Frequent, 1½ miles east of Elk Ranch Reservoir.
2. **E. cernuum** Nutt. Pothole area south of Signal Mt.
3. **E. flavum** Nutt. Not seen in G.T.N.P., but reported in Y.N.P.
4. **E. heracleoides** Nutt. Frequent at Colter Bay Village.
5. **E. ovalifolium** Nutt. var. **depressum** Blank. Frequent in alpine regions above 10,000 ft.
6. **E. umbellatum** Torr. A highly variable species with several varieties. Our most common var. **subalpinum** (Greene) Jones is abundant on the outwash and moraines throughout the valley.

OXYRIA Mountain Sorrel

Perennial herbs. Lvs mostly basal, reniform blades, sheathing stipules. Fls perfect, paniculate; perianth 4-parted; stamens 6; pistil with 2 carpels; fruit lenticular, winged.

1. **O. digyna** (L.) Hill. Frequent on talus slopes of the major canyons from 8,500 to 10,000 ft., circumboreal in N. Am.

POLYGONUM Knotweed; Smartweed

Annual or perennial herbs or subshrubs; stipules usually sheathing. Lvs alt, usually entire and petiolate. Fls perfect to unisexual, generally in spikelike racemes; perianth mostly 5-parted; stamens generally 8; pistil usually with 3 carpels; achenes lenticular to 3 angled.

1 Plants annual, generally in dry soil, but sometimes on wet soil; usually with a taproot

2 Stems twining; sagittate leaf base .... **P. convolvulus**

2 Stems not twining; leaves not sagittate

3 Pedicels becoming sharply recurved, older flowers reflexed .... **P. douglasii**

3 Pedicels erect or spreading but not reflexed

4 Flowers in spike-like racemes; leaves linear to linear-lanceolate
5 Flower bracts usually with a white margin; plants often greater than 7 cm .. **P. confertiflorum**
5 Flower bracts only slightly white margined; plants usually less than 7 cm .... **P. kelloggii**
4 Flowers usually axillary to leaves; leaves various
6 Achenes black, shining and smooth ... **P. minimum**
6 Achenes brown to yellow-green, not shining, often roughened .... **P. aviculare**
1 Plants perennial, usually growing in or near water; often with rhizomes
7 Flowers whitish, in a terminal head or spicate raceme; leaves mostly basal
8 Plants having lower flowers replaced by bulblets .... **P. viviparum**
8 Plants lacking bulblets .... **P. bistortoides**
7 Flowers pink, in axillary and terminal racemes on leafy stems
9 Inflorescences usually 4 cm long, peduncles usually glandular-pubescent .... **P. coccineum**
9 Inflorescences usually less than 4 cm long, peduncles generally glabrous .... **P. amphibium**

1. **P. amphibium** L. Water Ladysthumb. Aquatic or semi-aquatic, locally common in north end of valley.
†2. **P. aviculare** L. Common in waste places of the valley.
3. **P. bistortoides** Pursh. Bistort. Common along streambanks and wet meadows in the valley and up to 10,000 ft.
4. **P. coccineum** Muhl. Locally common in shallow pond one mile north of Colter Bay Village.
5. **P. confertiflorum** Nutt. Not seen in G.T.N.P. but reported in Y.N.P.
†6. **P. convolvulus** L. European weed on waste or cultivated land.
7. **P. douglasii** Greene. Locally frequent in rocks above Graul's Mine, 3 miles up Webb Canyon; also in disturbed sites of the valley.
8. **P. kelloggii** Greene. Dry pool near Swan Lake.
9. **P. minimum** Wats. Infrequent in rocks on slopes of Symmetry Spire, 7,200 ft. also Webb canyon.
10. **P. viviparum** L. Infrequent in meadows and streambanks of some canyons. Vivaparous.

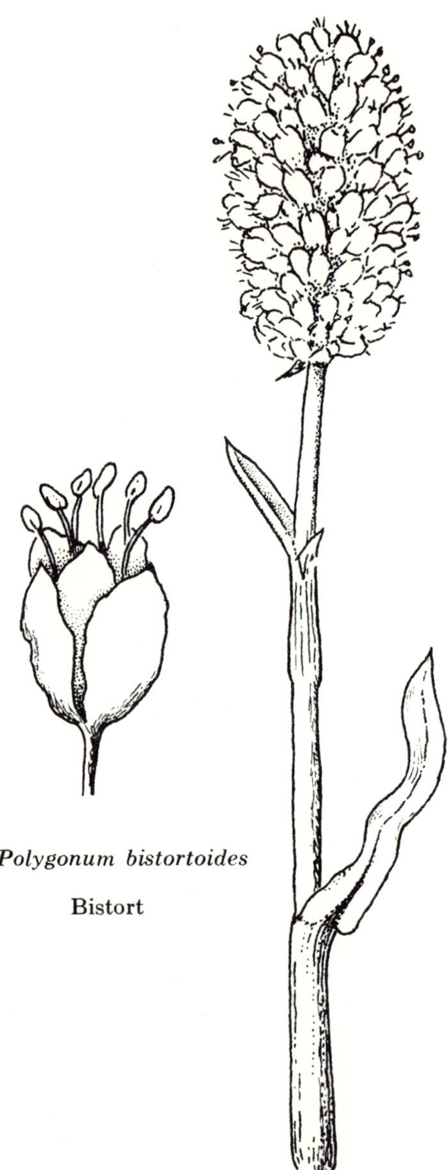

Polygonum bistortoides

Bistort

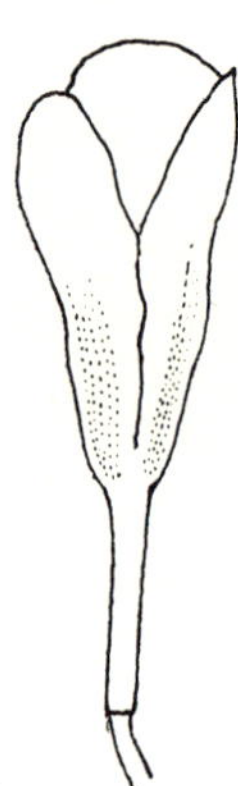

Perianth abruptly stiplelike

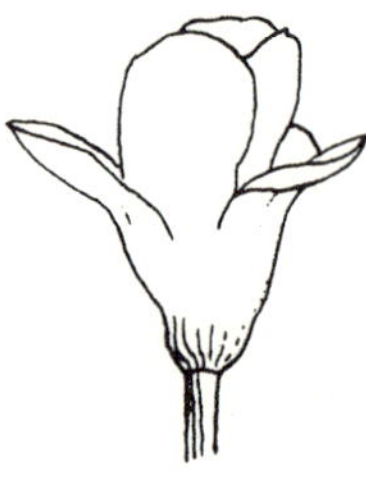

Perianth not abruptly stipelike

Eriognum

Polygonaceae

RUMEX Dock; Sorrel

Annual, biennial or perennial herbs, frequently with red surfaces. Lvs mostly alt, petioled. Fls small in large panicles, mostly perfect but sometimes unisexual; perianth 6-parted; stamens 6; styles 3; achenes 3-angled, brownish to nearly black.

1 Plants mostly dioecious; leaves either hastate or sagittate, or elliptic to oblanceolate
  2 Leaf blades more or less hastate to sagittate; weedy species .... **R. acetosella**
  2 Leaf blades never hastate or sagittate; plants of the mountains .... **R. paucifolius**
1 Plants with perfect flowers; leaves of various shapes but never hastate
  3 Plants with rhizomes; leaves often leathery .....**R. venosus**
  3 Plants with a strong taproot, rhizomes lacking; leaves various
    4 Inner perianth segments becoming more or less strongly toothed or dissected .... **R. maritimus**
    4 Inner perianth segments entire to slightly erose or denticulate
      5 Stems erect but lacking axillary shoots; leaf-margins crisp and wavy .... **R. crispus**
      5 Stems erect with well-developed axillary shoots; leaf-margins not wavy .... **R. salicifolius**

†1. **R. acetosella** L. European weed with tiny, reddish or yellowish flowers in slender loose spikes; disturbed sites.
†2. **R. crispus** L. European weed, common in disturbed sites.
3. **R. maritimus** L. [*R. fueginus* Phil.]. Frequent in wet places along the Snake River.
4. **R. paucifolius** Nutt. Common in meadows and open slopes of the valley.
5. **R. salicifolius** Weinm. ssp. **triangulivalvis** Danser. Frequent along roadsides throughout the valley.
6. **R. venosus** Pursh. Not seen in G.T.N.P., but known to be in adjacent areas. Large fruits and fleshy leaves.

PORTULACACEAE Purslane Family

Herbs/shrubs, often fleshy. Lvs alt/opp/basal, simple, stip. Infl various. Fls bisex. reg, ov usually superior. Ca 2, Co 3-∞,

S 3-∞ epipetalous and opp, P 2-3*, free central placentation. Capsule.

1 Flowers numerous in head-like clusters; sepals 2, with scarious margin; petals 4 .... **Spraguea**
1 Flowers single to many, usually in open racemes or cymes; sepals 2-many, usually not scarious; petals mostly 5 or more
  2 Petals and stamens usually 5; capsule dehiscent from the summit by 3 valves .... **Claytonia**
  2 Petals and stamens often more than 5; capsule circumscissile near the base .... **Lewisia**

CLAYTONIA Springbeauty

Perennial herbs with fleshy roots or corms, 1-several flowering stems. Lvs opp. Fls usually several, racemose; sepals 2, persistent; petals generally 5, white to pink or rose; stamens 5; capsule dehiscent from the top.

1 Plants from a globose corm; basal leaves 1-few ... **C. lanceolata**
1 Plants from a heavy thickened taproot; basal leaves numerous .... **C. megarhiza**

1. C. lanceolata Pursh. Common in sagebrush community to alpine slopes.
Stem leaves narrowly lanceolate; stems many ....var. **multiscapa** (Rydb.) C. L. Hitchc.
Stem leaves broader; stems 1-several ...... var. **lanceolata**

2. C. megarhiza (Gray) Parry. Rare on talus slopes, on Mt. Moran and Buck Mt. above 10,000 ft.

LEWISIA Lewisia

Perennial herbs with fleshy or corm-like roots. Lvs mostly basal. Fls single or many; sepals 2 or 4-9; petals 5-18, white to pink or magenta; stamens 5-many; capsule circumsissile at base.

1 Plants arising from a globose corm; no basal leaves at time of flowering .... **L. triphylla**
1 Plants arising from turnip or carrot-shaped root; basal leaves present at flowering
  2 Sepals 5-9 ....**L. rediviva**
  2 Sepals 2 ....**L. pygmaea**

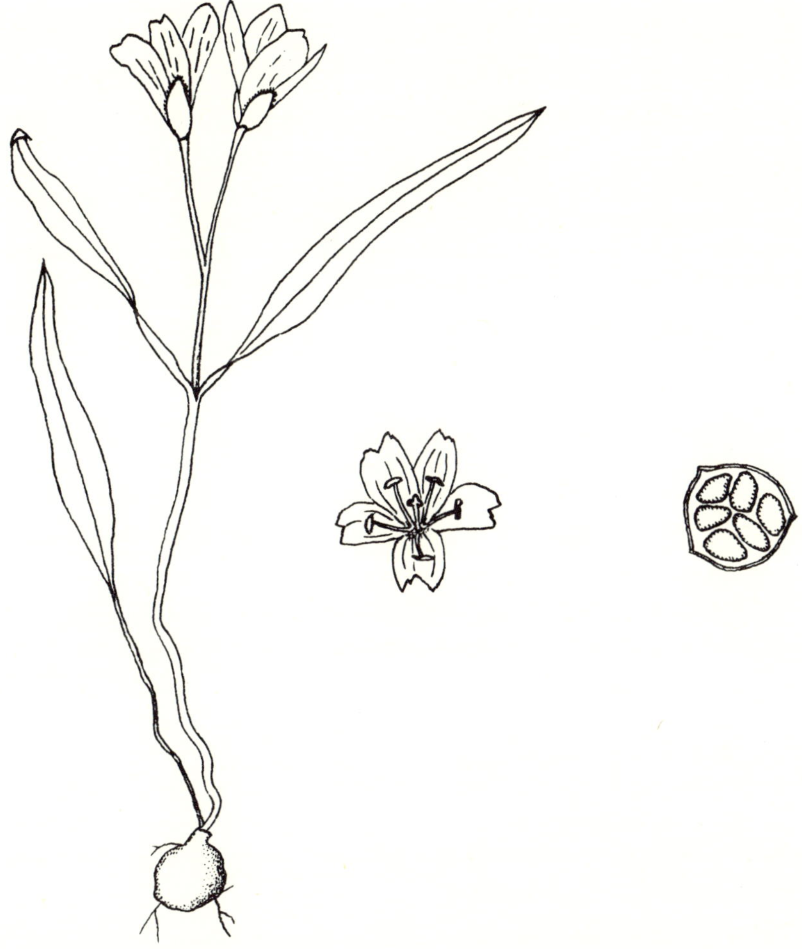

Claytonia lanceolata **Springbeauty** **Portulacaceae**

1. **L. pygmaea** (Gray) Robins. var. **pygmaea.** Locally frequent in sagebrush north of Timbered Island, also on Table Mt.
2. **L. rediviva** Pursh. Bitteroot. Not seen in G.T.N.P. but known to be in Y.N.P. in vicinity of Mammoth.
3. **L. triphylla** (Wats.) Robins. Infrequent in wet moss ¼ mile below Hidden Falls, also in Alaska Basin.

SPRAGUEA Pussypaws

Fleshy herbs. Lvs basal. Fls capitate; sepals with paper-like margins, white to pink, increasing in size with age, much larger than petals; capsule flattened; seeds black and shiny.

1. **S. umbellata** Torr. var. **umbellata.** Infrequent along Skyline Trail south of Static Peak.

PRIMULACEAE Primrose Family

Herbs. Lvs alt/opp/basal, simple, no stip. Infl various. Fls bisex, reg, ov superior (in ours). Ca 5*, Co 5*, S 5, epipetalous and opp, P 5*, stigma capitate, free central placentation. Capsule.

1 Corolla lobes several times longer than the corolla tube, abruptly reflexed; stamens extending beyond corolla lobes .... **Dodecatheon**

1. Corolla lobes less than twice as long as the corolla tube, not reflexed; stamens usually within the corolla tube
   2 Flowers usually less than 6 mm long, white ..... **Androsace**
   2 Flowers usually over 6 mm long, pink to lavender .. **Primula**

ANDROSACE Rock-jasmine

Annual or perennial herbs. Lvs in basal rosette, simple. Fls in umbel with involucral bracts; corolla white, tubular campanulate, lobes shorter than the tube; stamens included; capsule membranous; seeds pitted.

1 Flowers less than 3 mm long; leaves with distinct petioles; seeds yellow .... **A. filiformis**
1 Flowers more than 3 mm long; leaves narrowed gradually to the base; seeds brown .... **A. septentrionalis**

1. **A. filiformis** Retz. Frequent on east shore of Jackson Lake and slopes of Signal Mt.

2. **A. septentrionalis** L. Infrequent along small streams. A variable species with respect to stature and pubescence.

DODECATHEON Shooting Star

Perennial herbs often with small bulblets among the roots. Lvs basal petiolate, entire to dentate. Fls in umbel with involucral bracts, 4-5-merous; corolla white to pink to purple, with strongly reflexed lobes; stamens converging around the style; capsule 1-loculed.

1 Stigma conspicuously capitate; filaments usually not greater than 1 mm long .... **D. jeffreyi**
1 Stigma not conspicuously capitate; filaments usually united into a tube over 1 mm long
  2 Filaments generally less than 1.5 mm; connectives cross-wrinkled; capsule opening with a small lid .... **D. conjugens**
  2 Filaments generally greater than 1.5 mm; connectives often either smooth or wrinkled lengthwise; capsule opening by valves .... **D. pulchellum**

1. **D. conjugens** Greene. Frequent in the sagebrush flats, especially around Jenny Lake.
2. **D. jeffreyi** van Houtte. Reported in Waterfalls Canyon.
3. **D. pulchellum** (Raf.) Merrill. Frequent on moist open slopes above 8,000 ft.

PRIMULA Primrose

Herbaceous perennials. Lvs in a basal rosette. Fls showy, umbellate; calyx 5-lobed persistent; corolla salverform, reddish-purple; stamens attached in the upper third of the corolla tube; capsule 1-loculed, 5 valves.

1. **P. parryi** Gray. Frequent along streams in the higher canyons and on mountain slopes above 9,000 ft. Flowers are ill-smelling.

PYROLACEAE Wintergreen Family

Herbs. Lvs opp/alt/whorled, simple, evergreen or much reduced. Infl racemose/clustered/fls solitary. Fls bisex, reg, ov superior. Ca 5, Co 5(*), S 10, P 4-5*. Capsule.

Primula parryi Parry's Primrose **Primulaceae**

1 Plants without green leaves; plants reddish brown or pink, parasitic on soil fungi
  2 Stems fleshy, 1-2 dm high, few flowered; petals distinct .... **Hypopitys**
  2 Stems firmly erect, 3-8 dm high, many flowered; petals united .... **Pterospora**
1 Plants with green leaves present
  3 Stems bearing whorls of leathery leaves with serrate margins; inflorescence almost umbellate .... **Chimaphila**
  3 Stems with mostly basal leaves; flowers racemose or single
    4 Flowers single and terminal .... **Moneses**
    4 Flowers two or more in racemes .... **Pyrola**

CHIMAPHILA Pipsissewa; Prince's Pine

Low semishrubs from horizontal rootstocks, slightly woody. Lvs leathery, persistent, short-petioled, serrate margins. Fls 2-several in small umbels or corymbs; sepals 5, petals 5, spreading, pink; stamens 10; capsule loculicidal.

1. **C. umbellata** (L.) Barton var. **occidentalis** (Rydb.) Blake. Frequent under conifers on the shores of the valley lakes.

HYPOPITYS Pinesap

Fleshy herbs without chlorophyll, growing as parasites on soil fungi. Lvs scalelike. Fls axillary in terminal racemes; perianth usually 4-merous; petals saccate at the base, usually hairy on both surfaces; stamens twice the number of petals; capsule 5-8 mm long. Plants 5-25 cm tall, pink to straw colored.

1. **H. monotropa** Crantz. [*Monotropa hypopitys* L.]. Rare in deep shade of coniferous forest. Seen only on the west side of Leigh Lake.

MONESES One-flowered Wintergreen; Woodnymph

Perennial herbs. Lvs crowded near ground, with short petioles, orbicular or ovate. Fls single on a scape, nodding; sepals ovate, ciliolate; petals white, spreading; stamens 10, stigma with 5 erect lobes; capsule nearly globose.

1. **M. uniflora** (L.) Gray. [*Pyrola uniflora* L.]. Rare around the shores of the valley lakes.

PTEROSPORA Pinedrops

Plants reddish brown, lacking cholorophyll, viscid-hairy. Lvs much reduced. Fls many in an elongated raceme; calyx 5 parted; corolla urn-shaped, yellowish brown; stamens 10; fruit a capsule.

1. **P. andromedea** Nutt. Frequent, deep humus in coniferous forests of the valley.

PYROLA Shinleaf; Wintergreen

Perennial herbs with slender rhizomes. Lvs basal, evergreen. Fls borne in several-flowered racemes on slender pedicels; calyx 5-parted; petals 5, distinct; stamens 10, anthers opening by pores; fruit a 5-lobed, loculicidal capsule.

1 Style straight or nearly so; flowers less than 1 cm broad
  2 Styles over 2 mm long; flowers of raceme directed to one side only; petals white .... **P. secunda**
  2 Styles less than 2 mm long; normal raceme; petals flesh colored or pinkish .... **P. minor**
1 Style curved; flowers usually at least 1 cm broad
  3 Leaves more or less white mottled along the veins .. **P. picta**
  3 Leaves uniformly green or at least not mottled
    4 Petals pinkish to rose-purple .... **P. asarifolia**
    4 Petals pale yellowish to greenish white
      5 Leaves mostly acute, seldom over 2.5 cm in width; racemes 10-30 flowered .... **P. dentata**
      5 Leaves mostly rounded at apex, often over 2.5 cm in width; racemes less than 10 flowered . **P. chlorantha**

1. **P. asarifolia** Michx. Frequent on moist ground in coniferous forests of the valley.
2. **P. chlorantha** Sw. [*P. virens* Schweigg]. Infrequent on moist soil, Bradley Creek and north slope of Signal Mountain.
3. **P. dentata** Smith. Rare on dry hillsides above Amphitheater Lake, 10,000 ft.
4. **P. minor** L. Rare in alpine area at head of Avalanche Canyon, also lower portion of Indian Paintbrush Canyon.
5. **P. picta** Smith. Coniferous forest, Death Canyon.
6. **P. secunda** L. Frequent under conifers around the shores of the valley lakes.

Pyrola secunda One sided Shinleaf **Pyrolaceae**

RANUNCULACEAE Buttercup Family

Herbs/woody vines. Lvs usually alt, simple/compound, usually no stip. Infl various. Fls bisex, reg, ov superior. Ca 3-∞, Co 3-∞, S ∞, P 1-∞. Achene/follicle/berry.

1 Flowers irregular, bilaterally symmetrical, conspicuous
  2 Petals 2, hidden by the sepals; upper sepal hooded .... **Aconitum**
  2 Petals 4, not hidden by the sepals; upper sepal extended into a spur .... **Delphinium**
1 Flowers mostly regular, radially symmetrical, often inconconspicuous
  3 Pistil 1; fruit a berry .... **Actaea**
  3 Pistil 2 or more; fruit achene or follicle
    4 Plants annual with linear leaves; flowers inconspicuous, sepals spurred .... **Myosurus**
    4 Plants mostly perennial without linear leaves; flowers mostly showy
      5 Petals prominently spurred, showy .... **Aquilegia**
      5 Petals not spurred, sometimes lacking
        6 Petals and sepals obviously different; petals with nectary pit near base (use hand lens) .... **Ranunculus**
        6 Petals lacking, sepals often petaloid
          7 Sepals small, greenish or white, less conspicuous than the stamens; leaves alternate .... **Thalictrum**
          7 Sepals large and showy, various colors, more conspicuous than the stamens; leaves opposite, alternate or whorled
            8 Sepals generally 4; styles long and plumose .... **Clematis**
            8 Sepals generally 5 or more; styles mostly short and not plumose
              9 Fruits achenes; upper stem leaves in a whorl below the inflorescence .... **Anemone**
              9 Fruits folliciles; leaves various
                10 Leaves simple, margins crenate or dentate .... **Caltha**
                10 Leaves deeply palmately lobed . **Trollius**

ACONITUM Monkshood

Perennial herbs. Lvs alt, palmately lobed. Fls borne in bracteate racemes; sepals 5, petaloid, mostly purple, upper one arched and hoodlike; petals 2, hidden within the hood; stamens many; pistils 3-5; 3-5 follicles.

1. **A. columbianum** Nutt. Locally common in moist woods and meadows, often along streams, in the valley and extending into the canyons up to 9,000 ft.

ACTAEA Baneberry

Perennial herbs. Lvs ternately compound. Fls small, racemose; sepals 3-5, petaloid, white; petals 5-10, white; stamens numerous; pistil 1; fruit a berry.

1. **A. rubra** (Ait.) Willd. Frequent in the major canyons and around the shores of the valley lakes.

ANEMONE Windflower; Anemone

Perennial herbs. Lvs palmately dissected, basal, except for a whorl of 3 simple or dissected involucral leaves below the flowers. Fls solitary and terminal on peduncles; sepals usually 5, yellowish white, red, blue or purple; petals lacking; stamens and pistils numerous; achenes glabrous to densely woolly.

1. **A. multifida** Poir. [*A. globosa* Nutt.].
   Stems 2-6 dm; frequent along the Snake River .... var. **multifida**
   Stems less than 2 dm; infrequent in the alpine tundra .... var. **tetonensis** (Porter) Hitchc.

AQUILEGIA Columbine

Perennial herbs. Lvs mostly basal, bi- or ternately compound. Fls showy, white to blue, yellow or red; sepals 5, petaloid; petals 5, forming a spur at the base; stamens numerous; pistils 5; follicles.

1 Sepals yellow, sometimes tinged with red .... **A. flavescens**
1 Sepals blue or purple to white .... **A. caerulea**

1. **A. caerulea** James. Common in major canyons and subalpine meadows.

2. **A. flavescens** Wats. Infrequent in moist mountain meadows to alpine slopes.

CALTHA Marshmarigold

Herbaceous perennials. Lvs simple, reniform to cordate-oblong, crenate or dentate margins. Fls showy, white (in ours); sepals 5-10, petaloid; stamens numerous; pistils 5-10; follicles.

1. **C. leptosepala** DC. Frequent in subalpine and alpine wet meadows above 8,500 ft. Fls push through melting snow.

CLEMATIS Virgins-bower; Clematis

Herbaceous perennials to woody vines. Lvs opp or whorled, 1-several times ternate or pinnate. Fls showy, apetalous; sepals mostly 4, petaloid, white to bluish or purplish; stamens numerous; achenes numerous, styles long becoming feathery.

1 Fls white, sepals less than 1 cm long .... **C. ligusticifolia**
1 Fls blue or purple, sepals more than 1 cm long
  2 Plants more or less woody and vine like; sepals spreading, not connate at base .... **C. occidentalis**
  2 Plants herbaceous; sepals erect, connate at base .... **C. hirsutissima**

1. **C. hirsutissima** Pursh. Infrequent in the sagebrush of the valley and in some of the higher canyons above 9,000 ft.
2. **C. ligusticifolia** Nutt. Infrequent on trail, 2-3 miles up Death Canyon, 7,500 ft.
3. **C. occidentalis** (Hornem.) DC. var. **grosseserrata** (Rydb.) Pringle. Infrequent in Cascade Canyon and along Pilgrim Creek.

DELPHINIUM Larkspur; Delphinium

Perennial herbs with fibrous to fleshy roots. Lvs alt, palmately parted or divided. Fls borne in simple to compound racemes, irregular, bluish or purplish (in ours); sepals 5, the upper one conspicuously spurred; petals generally 4, smaller than the sepals; stamens numerous; follicles generally 3. A critical genus treated in different ways.

1 Plants 10-30 dm tall; stems hollow .... **D. occidentale**
1 Plants less than 10 dm tall; stems solid

2 Lower petals shallowly notched; sepals unequal, lower pair longer .... **D. bicolor**
2 Lower petals deeply notched; sepals nearly equal .... **D. nuttallianum**

1. **D. bicolor** Nutt. Reported in Yellowstone and often confused with *D. nuttallianum.* Grassland and open forest.
2. **D. nuttallianum** Pritz. Common in gravelly soil throughout the valley,
3. **D. occidentale** Wats. Frequent in moist soil in open woods.

MYOSURUS Mouse-tail

Acaulescent annual herbs. Lvs linear and basal. Fls solitary on leafless scapes; sepals 5, greenish, spurred at base; petals 5, linear; stamens 5 or 10; achenes numerous on elongated receptacle.

1. **M. minimus** L. Frequent in moist areas, especially in vernal pools along the Snake River.

RANUNCULUS Buttercup; Crowfoot; Water-buttercup

Generally perennial herbs with erect, creeping or floating stems. Lvs alt, simple to compound. Fls regular, often showy; sepals generally 5; petals mostly 5; commonly yellow, but sometimes white, nectar gland at the base; stamens numerous; achenes 5-numerous.

1 Flowers white .... **R. aquatilis**
1 Flowers yellow
2 Leaves entire to serrate, never dissected or compound, sometimes lobed at the tip; achenes glabrous
3 Achenes 50 or more, arranged in a columnar group, ribbed lengthwise; leaves cordate to rhombic .... **R. cymbalaria**
3 Achenes less than 50, not ribbed lengthwise; leaves linear .... **R. flammula**
2 Leaves usually deeply lobed to compound; achenes sometimes hairy or spiny
4 Plants aquatic or semi-aquatic; stems generally creeping and often rooting at the nodes; achenes glabrous
5 Plants annual; stems erect, no rooting at the nodes .... **R. sceleratus**

5 Plants perennial; stems mostly floating or reclining, rooting at the node
6 Petals 2-4.5 mm long; leaf lobes entire or notched; nectary scale adnate on the sides .... **R. natans**
6 Petals 3-10 mm long; leaves more or less finely dissected; nectary scale adnate only at base .... **R. flabellaris**
4 Plants not aquatic; stems mostly without rooting at the nodes; achenes either hairy or with conspicuous beak
7 Nectary scale elongate, the margins free for at least ½ of the length; achenes strongly compressed, never pubescent
8 Petals 9-20 mm long; basal leaves 5-7 pinnately lobed .... **R. orthorhynchus**
8 Petals usually less than 10 mm long; basal leaves rarely pinnately lobed
9 Petals much longer than sepals; beak of achene strongly recurved .... **R. acriformis**
9 Petals slightly if at all longer than the sepals; beak of achene not strongly recurved .... **R. macounii**
7 Nectary scale forming a small pocket; achenes only moderately compressed, sometimes pubescent
10 Petals 8-15 mm long, nectary scale 1.5-3 mm long, ciliate on upper margin; stem usually less than 15 cm, ascending to decumbent .... **R. glaberrimus**
10 Petals often less than 6 mm long, nectary scale often less than 1 mm, or broader than long; stems generally erect, often more than 15 cm long
11 Stems mostly less than 10 cm; roots fleshy, enlarged toward the tip, shaped like a baseball bat .... **R. jovis**
11 Stems mostly greater than 10 cm; roots tapering gradually to the tip
12 Basal leaf blades cuneate .... **R. inamoenus**
12 Basal leaves lobed or divided, often cordate .... **R. eschscholtzii**

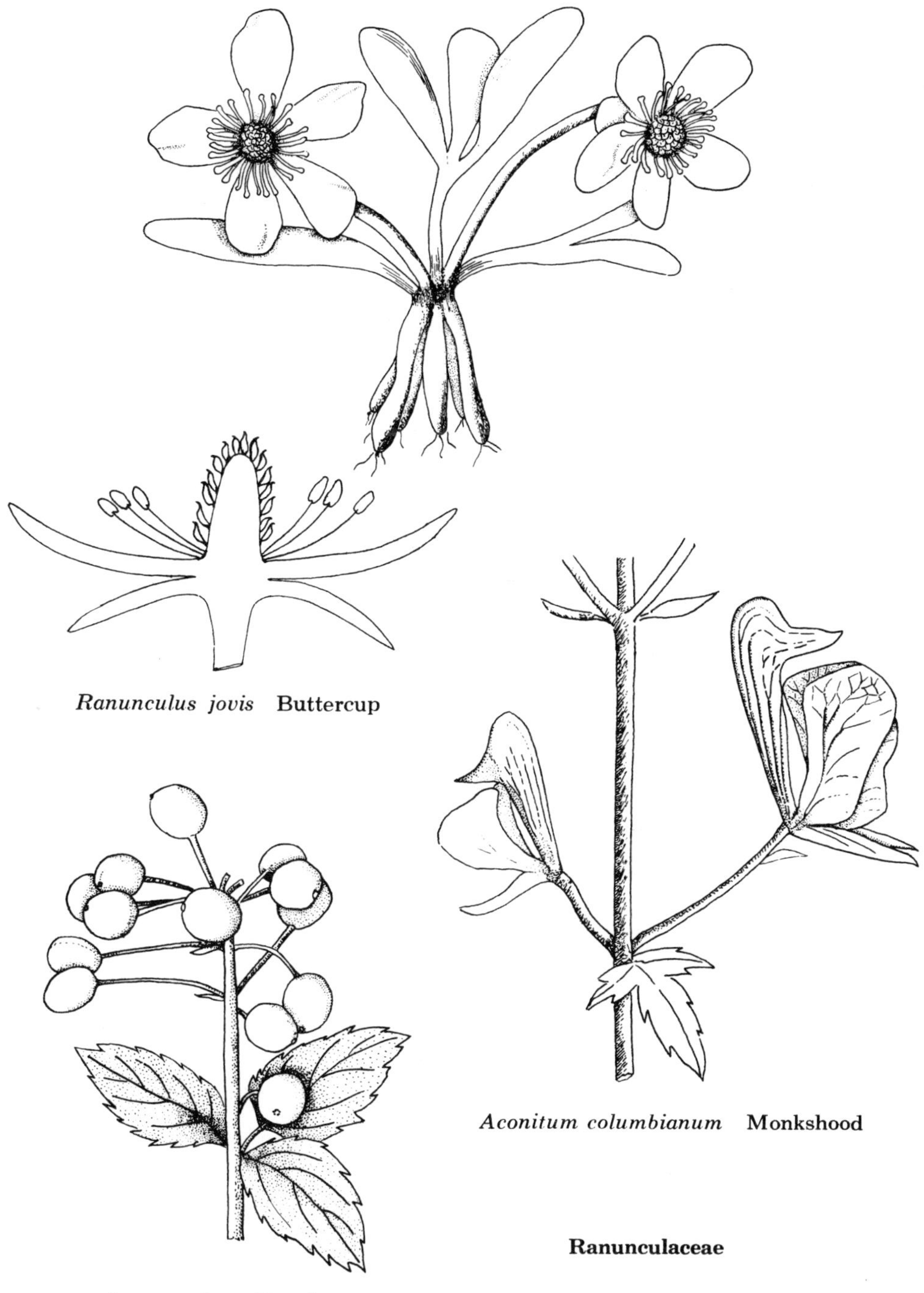

Ranunculus jovis Buttercup

Aconitum columbianum Monkshood

Actaea rubra Baneberry

Ranunculaceae

1. **R. acriformis** Gray. Frequent in wet places north and east of Jackson Lake Dam.
2. **R. aquatilis** L. var. **capillaceus** (Thuill.) DC. Frequent in pools and sluggish streams throughout the valley.
3. **R. cymbalaria** Pursh. Common in wet places along the Snake River.
4. **R. eschscholtzii** Schlecht. A widespread variable species in mountain meadows and talus slopes above 9,000 ft. with the following varieties:

   1 Main (middle) segment of the basal leaves usually twice-lobed into linear segments .... var. **alpinus** (Wats.) C. L. Hitchc.

   1 Main (middle) segment of the basal leaves no more than once-lobed

      2 Leaf segments mostly acute .... var. **suksdorfii** (Gray) Benson

      2 Leaf segments mostly round-tipped ... var. **eschscholtzii**

5. **R. flabellaris** Raf. Infrequent in pond north of Berry Creek Cabin.
6. **R. flammula** L. var. **filiformis** (Michx.) Hook. Frequent at edge of shallow ponds north of Colter Bay Village.
7. **R. glaberrimus** Hook. Frequent in open forest in area of the Oxbow Bend in the Snake River.
8. **R. inamoenus** Greene. Frequent in sedge meadow of the Oxbow Bend.
9. **R. jovis** A. Nels. Common in the early spring on the sagebrush flats of the valley floor.
10. **R. macounii** Britt. Irrigated meadows on the Elk Ranch.
11. **R. natans** C. A. Meyer. Common in wet places near the Oxbow Bend.
12. **R. orthorhynchus** Hook. Reported in area of Trail Creek and Berry Creek.
13. **R. sceleratus** L. var. **multifidus** Nutt. Common in wet places near the Oxbow Bend.

THALICTRUM Meadowrue

Perennial herbs. Lvs alt, bi- to tri-ternately compound. Fls apetalous, perfect to dioecious, borne in racemes or panicles; sepals 4 or 5, greenish white; stamens numerous; pistils 4-9; achenes sessile or stipitate.

1 Flowers perfect; anthers about 0.7 mm long .. **T. sparsiflorum**

1 Flowers at least in part imperfect; anthers at least 1 mm long
  2 Achenes spreading to reflexed, elliptic to spindle shaped; stigma often purple, 3-4.5 mm long, including the style .... **T. occidentale**
  2 Achenes usually ascending or erect, obliquely elliptic in outline; stigma rarely purple, 1.5-3.5 mm long, including the style .... **T. fendleri**

1. **T. fendleri** Engelm. Frequent along streams in the coniferous forest of the valley.
2. **T. occidentale** Gray. Frequent in wet sites in the area of the Oxbow Bend of the Snake River.
3. **T. sparsiflorum** Turcz. Infrequent in damp woods, west shore of Jenny Lake.

TROLLIUS Globeflower

Perennial herbs. Lvs alt, palmately lobed to compound. Fls solitary on terminal peduncles, apetalous; sepals 5-12, petaloid, white to yellow; stamens numerous; follicles 8-11 mm long.

1. **T. laxus** Salisb. Frequent in wet meadows of subalpine and alpine areas of Teton Range.

RHAMNACEAE Buckthorn Family

Woody. Lvs usually alt, simple, stip. Infl corymbose/cymose/fasciculate. Fls uni/bisex, reg, ov superior/inferior, disc usually present. Ca 4-5, Co 4-5, S 4-5, P 2-4*. Capsule/berry-like drupe.

1 Flowers white; petals with a distinct claw; fruit green or brown .... **Ceanothus**
1 Flowers greenish; petals generally lacking; fruit bluish-black .... **Rhamnus**

CEANOTHUS Ceanothus; Buckbrush

Small shrubs (in ours). Lvs alt. Fls in panicles or umbels; 5-merous; fruit splitting into 3 1-seeded carpels.

1. **C. velutinus** Dougl. var. **velutinus.** Frequent around the shores of the valley lakes.

RHAMNUS Buckthorn

Shrubs (in ours). Lvs alt, simple with obvious veins. Fls borne in axillary clusters; calyx 4-5 lobed; petals absent; berry-like fruit.

1. R. alnifolia L'Her. Frequent along the Snake River and the shore of Jackson Lake.

ROSACEAE Rose Family

Herbs/woody. Lvs usually alt, simple/compound often serrate, stip. Infl various. Fls usually bisex, reg, ov superior/inferior. Ca 5, Co 5, S ∞, P 1-∞*. Follicle/achene/drupe/pome.

1 Ovary inferior, fused with the fleshy hypanthium; fruit a pome (apple-like)
  2 Leaves pinnate; flowers numerous .... **Sorbus**
  2 Leaves simple; flowers 1 to few
    3 Plants possessing stout thorns; carpels stony at maturity .... **Crataegus**
    3 Plants lacking thorns; carpels have thin walls .... **Amelanchier**
1 Ovary superior, not fused with hypanthium; fruits achenes, follicles, or several fused drupelets
  4 Carpel solitary; fruit a fleshy, 1-seeded drupe (cherry-like) .... **Prunus**
  4 Carpels more than one; styles 1 to many; fruit of follicles, achenes, or drupelets
    5 Plants herbaceous or dwarf matted undershrubs
      6 Styles persistent on achenes, usually elongating and plumose or hooked at maturity
        7 Plants dwarf shrubs; flowers solitary .... **Dryas**
        7 Plants herbs; flowers usually in cymes ...... **Geum**
      6 Styles deciduous, fruits sometimes follicles
        8 Stamens numerous, more than 10
          9 Receptacle becoming large and pulpy at maturity, edible; flowers white .... **Fragaria**
          9 Receptacle not becoming large and pulpy; flowers yellow, white or maroon
            10 Plants densely caespitose, fruits follicles .... **Petrophytum**

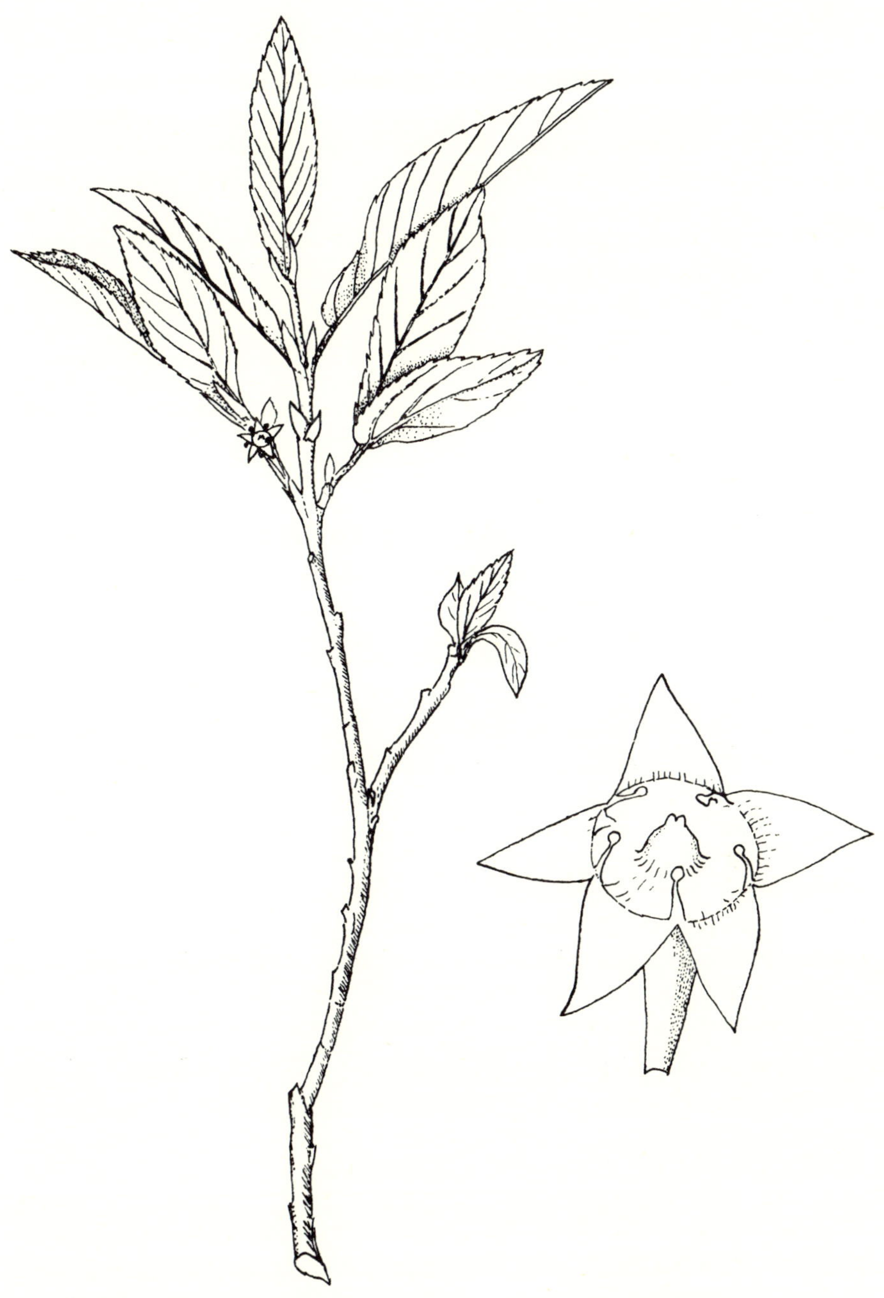

Rhamnus alnifolia Buckthorn **Rhamnaceae**

10 Plants not densely caespitose, fruits achenes .... **Potentilla**
8 Stamens 5 or 10, occasionally more
11 Leaves pinnate .... **Ivesia**
11 Leaves trifoliate .... **Sibbaldia**
5 Plants obvious shrubs or trees
12 Leaves simple
13 Leaves palmately lobed
14 Flowers in corymbs; bark peeling off in strips .... **Physocarpus**
14 Flowers in panicles; bark not peeling .... **Rubus**
13 Leaves various but not palmately lobed
15 Leaves evergreen .... **Cercocarpus**
15 Leaves deciduous
16 Leaves apically 3-cleft; flowers yellow **Purshia**
16 Leaves simply toothed; flowers pink or white .... **Spiraea**
12 Leaves compound
17 Plants armed with prickles
18 Flowers usually white; fruit of several fleshy drupelets .... **Rubus**
18 Flowers pink to deep rose; achenes enclosed in fleshy hypanthium .... **Rosa**
17 Plants unarmed; flowers yellow .... **Potentilla**

AMELANCHIER Serviceberry

Unarmed shrubs or small trees. Lvs alt, simple, often with serrate margin, small stipules. Fls borne in small racemes; petals 5, white, linear to oblanceolate; stamens variable in number; styles 2-5; ovary inferior; fruit a small berry-like pome, reddish to purple.

1. **A. alnifolia** Nutt. Common along the Snake River and around the shores of the valley lakes. Fruits are edible.

CERCOCARPUS Mountain Mahogany

Shrubs or small trees. Lvs alt, simple, entire to toothed margins. Fls small 1-several in axillary clusters; calyx persistent; petals none; stamens numerous; pistil 1, forming a linear achene, with the style becoming twisted and plumose.

1. **C. ledifolius** Nutt. Infrequent in the lower portions of Granite Canyon, and common in Darby Canyon.

CRATAEGUS Hawthorn

Shrubs or small trees, usually well armed with strong thorns. Lvs alt, simple, petiolate, toothed to lobed. Fls in axillary corymbs; petals 5, white; stamens 10-15; pistil with 2-5 carpels; ovary inferior; fruit a pome, red, purple, or black.

1. **C. douglasii** Lindl. var **douglasii.** Frequent along the Snake River and the Taggart Lake Trail.

DRYAS Mountain Avens

Dwarf, matted undershrubs. Lvs alt, simple, crenate or entire. Fls solitary on naked peduncles; petals usually 8-10, white to yellow; pistils numerous; the persistent styles elongating and plumose in fruit; fruit an achene.

1. **D. octopetala** L. var. **angustifolia** C. L. Hitchc. Infrequent in rocky soil above 10,000 ft., Skyline Trail.

FRAGARIA Strawberry

Perennial herbs, spreading by runners. Lvs basal, trifoliate, the leaflets coarsely toothed. Fls showy, several to a scape; petals 5, white; stamens about 20, borne at edge of the hypanthium; pistils numerous; receptacle hemispheric or conic, enlarged and juicy at maturity; achenes rather small. A difficult genus.

1 Leaves relatively thin and more or less veiny and somewhat pilose-silky, not glaucous .... **F. vesca**
1 Leaves thick and not prominently veined and nearly always glabrous, glaucous .... **F. virginiana**

1. **F. vesca** L. var. **bracteata** (Heller) Davis. Frequent on the moraines of the valley floor.
2. **F. virginiana** Duchn. var. **glauca** S. Wats. Frequent in vicinity of the Oxbow Bend of the Snake River.

GEUM Avens

Perennial rhizomatous herbs. Lvs alt, opp, or basal, cauline leaves mostly 3-lobed. Fls in cymes; hypanthium free of pistils; petals 5, usually yellow; stamens numerous; pistils numerous; the achenes tipped by persistent long styles.

1 Sepals bent downward (reflexed) at anthesis; styles strongly bent and jointed .... **G. macrophyllum**

1 Sepals ascending or erect; styles neither bent nor jointed
  2 Petals pinkish-white; flowers nodding; a valley plant .... **G. triflorum**
  2 Petals yellow; flowers erect; above timber line ..... **G. rossii**

1. **G. macrophyllum** Willd. var. **perincisum** (Rydb.) Raup. Frequent in moist sties on the moraines of the valley.
2. **G. rossii** (R. Br.) Ser. var. **turbinatum** (Rydb.) C. L. Hitchc. Frequent on talus slopes above 9,500 ft.
3. **G. triflorum** Pursh var. **ciliatum** (Pursh) Fassett. Frequent in meadows and sagebrush flats of the valley.

IVESIA

Perennial herbs. Lvs mostly basal and pinnately compound. Fls borne in crowded cymes; hypanthium campanulate to saucer-shaped; petals yellow; stamens 5-20; pistils mostly 2-6; fruit an achene.

1. **I. gordonii** (Hook.) T. & G. Frequent on talus slopes above 10,000 ft. and occasionally along flood plains and stream banks in the valley.

PETROPHYTUM Rock Spirea

Depressed, woody shrubs. Lvs alt, crowded and tufted, simple, entire. Fls small borne in spike-like racemes, petals 5, white; stamens 20-40; pistils 3-7; fruit a follicle with 1 or 2 seeds.

1. **P. caespitosum** (Nutt.) Rydb. Infrequent on limestone ledges, north end of Blacktail Butte, 2 miles up Berry Creek, and Snake River Canyon on ledges above the river.

PHYSOCARPUS Ninebark

Shrubs with bark coming off in layers. Lvs alt, simple, usually palmately 3-5 lobed. Fls in corymbs; petals 5, white; stamens 20-40; pistils 1-5; fruit follicular.

†1. **P. malvaceus** (Greene) Kuntze. Recent introduction planted at the Jackson Lake Lodge, but may be naturally occurring in Snake River Canyon.

POTENTILLA Cinquefoil

Annual to perennial herbs or shrubs. Lvs alt, pinnately to digitately compound, leaflets toothed or dissected. Fls in cymes;

petals generally yellow, sometimes white (one species reddish or purple); stamens 10-30; pistils numerous; achenes smooth to reticulate. A difficult genus because of apomictic reproduction and hybridization between species.

1 Plants shrubs; achenes covered with hairs .... **P. fruticosa**
1 Plants herbs; achenes glabrous
  2 Flowers purple to deep red .... **P. palustris**
  2 Flowers white, cream or yellow
    3 Plants annual or biennial; without rootstocks **P. norvegica**
    3 Plants perennial; usually with rootstocks
      4 Stems creeping and rooting; flowers solitary ... . **P. anserina**
      4 Stems erect or ascending, not rooting; flowers in cymes
        5 Leaves pinnate, with 5-11 leaflets; style attached somewhat below midlength of the ovary
          6 Cymes narrow; plants mostly over 4 dm tall ....**P. arguta**
          6 Cymes usually open to diffuse; plants mostly less than 4 dm tall .... **P. glandulosa**
        5 Leaves palmate or if pinnate often with more than 11 leaflets; style attached near the top of the ovary
          7 Basal leaves usually odd-pinnate, some with 5 or more leaflets, occasionally a few ternate
            8 Plants bearing glandular hairs (use hand lens), 5-10 cm tall .... **P. brevifolia**
            8 Plants predominantly bearing hairs without glands; usually well over 10 cm tall
              9 Plants 1.5-4.5 dm tall; leaflets usually greenish on both surfaces (never tomentose) .... **P. diversifolia**
              9 Plants often less than 1.5 dm tall; leaflets tomentose at least beneath .... **P. concinna**
          7 Basal leaves predominantly palmately compound, sometimes with only 3 leaflets

10 Leaflets of basal leaves mostly 3, rarely 5
11 Leaves thin, glabrous or thinly pubescent, greenish on both surfaces .... **P. flabellifolia**
11 Leaves mostly thick and leathery, lower surface grayish with pubescence .... **P. nivea**
10 Leaflets of basal leaves mostly 5 or more
12 Plants spreading mostly less than 1 dm tall; leaves whitish tomentose beneath .... **P. concinna**
12 Plants averaging over 2 dm tall; leaves not whitish tomentose beneath, but often paler on the lower than on upper surface
13 Plants alpine; 1.5-4.5 dm tall .... **P. diversifolia**
13 Plants of the valley; up to 8 dm tall .... **P. gracilis**

1. **P. anserina** L. Frequent on mud flats, north end of Jackson Lake, also on a few drying ponds.
2. **P. arguta** Pursh var. **convallaria** (Rydb.) Th. Wolf. Frequent in meadows near the Oxbow Bend of the Snake River.
3. **P. brevifolia** Nutt. Infrequent in alpine areas such as Timberline Lake and Paintbrush-Leigh Canyon Divide.
4. **P. concinna** Rich var. **rubripes** (Rydb.) C. L. Hitchc. Infrequent on Static Peak and Table Mt., above 9,500 ft.
5. **P. diversifolia** Lehm. Common in alpine meadows and on ledges and rocky slopes.
6. **P. flabellifolia** Hook. Common in meadows of Holly Lake Cirque.
7. **P. fruticosa** L. Common along Snake River and frequent in the canyons.
8. **P. glandulosa** Lindl. var. **pseudorupestris** (Rydb.) Breit. Frequent in major canyons above 7,500 ft. This species varies in practically all floral and vegetative characteristics.
9. **P. gracilis** Dougl. Common in moist areas of the valley.
P. gracilis is the most variable species in the genus.
Leaflets dissected at least ⅔ of the way to the midvein .... var. **elmeri** (Rydb.) Jeps.

Leaflets dissected no more than ⅔ of the way to the midvein .... var. **glabrata** (Lehm.) C. L. Hitchc.

10. **P. nivea** L. Frequent on Static Peak, south facing slope.
11. **P. norvegica** L. Infrequent along the Snake River.
12. **P. palustris** (L.) Scop. Locally common at water's edge, Arizona Lake and Bradley Lake.

PRUNUS Plum; Cherry

Trees or shrubs. Lvs alt, simple, serrate. Fls in terminal racemes or corymbs; petals white; stamens 20-30; pistil 1, simple; fruit a drupe.

1. **P. virginiana** L. var. **melanocarpa** (A. Nels.) Sarg. Chokecherry. Common along the Snake River and on the moraines of the valley.

PURSHIA Bitterbrush

Shrubs 0.5-3 m tall. Lvs alt, deciduous, apically 3-cleft, margins revolute, white tomentose beneath. Fls solitary; petals 5, yellow; stamens usually 25; fruit an achene.

1. **P. tridentata** (Pursh) DC. Common in the sagebrush flats of the valley, particularly south of Moose. This is considered to be one of the best browse plants in western U.S.

ROSA Rose

Shrubs usually prickly. Lvs alt, odd-pinnately compound; stipules well developed. Fls strongly perigynous, borne singly or in cymes; petals 5, light pink to deep rose; pistils several; hypanthium fleshy, enclosing the achenes, commonly called a hip.

1. **R. woodsii** Lindl. var. **ultramontana** (Wats.) Jeps. Common along Snake River and its tributaries.

RUBUS Raspberry; Blackberry

Perennial shrubs with trailing to erect stems, mostly with prickles or bristles. Lvs alt, simple to ternate or pinnate, mostly with stipules. Fls often showy, perigynous, single to several; petals 5, white to red; stamens numerous; pistils several to many; aggregate fruit of weakly coherent drupelets.

1 Stems not prickly; leaves simple .... **R. parviflorus**

1 Stems prickly; leaves mostly compound

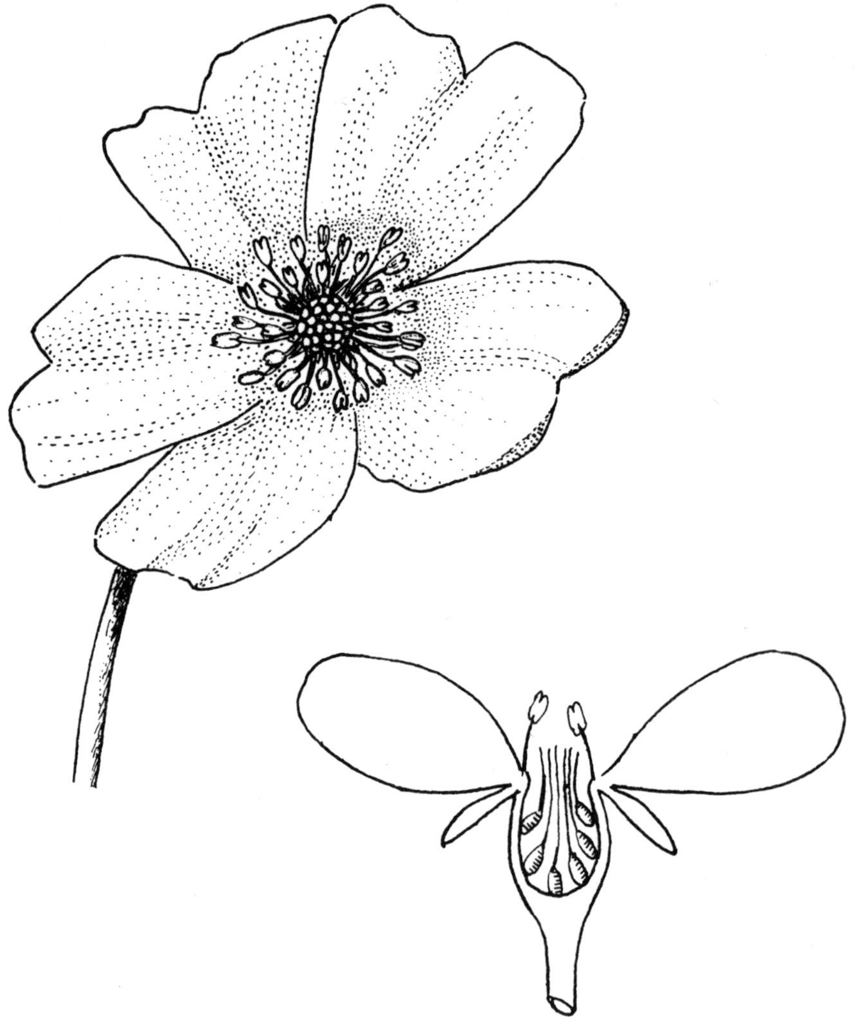

Rosa woodsii Wild Rose **Rosaceae**

2 Receptacle fleshy, forming part of the ripe fruit; prickles flattened or hooked .... **R. laciniatus**

2 Receptacle dry, usually not forming part of the ripe fruit; prickles neither flattened or hooked .... **R. idaeus**

1. **R. idaeus** L. ssp. **sachalinensis** (Levl.) Focke. Infrequent along Snake River, also frequent on talus slopes of the major canyons.
2. **R. laciniatus** Willd. Gravelly bank of Snake River near site of old town of Moran; a possible escaped cultigen.
3. **R. parviflorus** Nutt. Thimbleberry. Common on the moraines and in the lower portions of the major canyons.

SIBBALDIA

Low, perennial, tufted herbs. Lvs stipulate, ternate. Fls small, perigynous, borne in small cymes; petals 5, pale yellow; stamens 5; pistils 5-15; fruit an achene.

1. **S. procumbens** L. Frequent on the shores of subalpine lakes above 9,000 ft.

SORBUS Mountain Ash

Trees or shrubs. Lvs alt, pinnately compound, serrate leaflets. Fls borne in compound corymbs; petals 5, white; stamens 15-20; ovary inferior; fruit pomaceous.

1. **S. scopulina** Greene var. **scopulina.** Common on the lower slopes of the mountains and along the shores of the valley lakes.

SPIRAEA Spiraea

Shrubs. Lvs alt, simple, toothed, without stipules. Fls perigynous, borne in dense corymbs or panicles; petals 5, white to pink (in ours); stamens numerous; pistils usually 5, forming small follicles.

1 Petals white; plants mostly glabrous .... **S. betulifolia**

1 Petals pink; inflorescence and the lower surface of the leaves often puberulent .... **S. densiflora**

1. **S. betulifolia** Pall. var. **lucida** (Dougl.) C. L. Hitchc. Frequent on the moraines and shores of the valley lakes.
2. **S. densiflora** Nutt. Frequent on mountain slopes and in major canyons up to 9,000 ft.

RUBIACEAE Madder Family

Herbs (in ours). Lvs opp, simple, stip. Infl cymose/capitate/ fls solitary. Fls bisex, reg, ov inferior. Ca 4-5, Co 4-5, S 4-5, P 2-4*. Capsule/berry.

1 Leaves opposite; corolla funnelform to salverform . . . **Kelloggia**
1 Leaves whorled; corolla rotate, the tube much shorter than the lobes . . . . **Galium**

GALIUM Bedstraw

Annual or perennial herbs. Lvs simple, entire, with stipules, opposite or appearing in whorls of 3-8. Fls perfect or unisexual, borne in cymose inflorescences; calyx lobes obsolete; corolla white, united, with 3-5 lobes; stamens 3-5; fruit dry and indehiscent forming 2 mericarps.

1 Plants annual
  2 Stem glabrous; 2-4 leaves in a whorl . . . . **G. bifolium**
  2 Stem covered with downward bending stiff hairs; 5-8 leaves in a whorl . . . . **G. aparine**
1 Plants perennial with rhizomes
  3 Mericarps covered with uncinate hairs (hooked at the tip) . . . . **G. triflorum**
  3 Mericarps without hooked hairs
    4 Flowers solitary or few in inconspicuous inflorescences; corollas mostly 3-lobed . . . . **G. trifidum**
    4 Flowers numerous in a much-branched inflorescence; corollas mostly 4-lobed . . . . **G. boreale**

1. **G. aparine** L. var. **echinospermum** (Wallr.) Farw. Frequent, a weedy species in a variety of habitats; circumpolar.
2. **G. bifolium** Wats. Frequent under the aspen, along the Snake River.
3. **G. boreale** L. Common on morainal soils of the valley; circumpolar.
4. **G. trifidum** L. Frequent in moist places, vicinity of the Oxbow Bend of the Snake River; circumpolar.
5. **G. triflorum** Michx. Rare in vicinity of Treasure Mt., Teton Canyon, also in Webb Canyon.

KELLOGIA

Perennial herbs with creeping rhizomes. Lvs opp, entire, sessile with small interpetiolar stipules. Fls borne in cymose inflorescences; perianth 4-5 merous; corolla about 5 mm long, funnelform to salverform, pink or whitish; fruit dry, separated into two 1-seeded mericarps covered with hooked bristles.

1. **K. galioides** Torr. Infrequent among rocks, east face of Symmetry Spire.

SALICACEAE Willow Family

Woody, dioecious. Lvs usually alt, simple, stip. Infl catkin. Fls usually unisex, ov superior. Per 0, S 1-∞, P 2-4*. Capsule, seeds with tuft of hairs.

1 Leaf buds covered by several overlapping scales, resinous; stamens 6-many; trees with drooping catkins .... **Populus**
1 Leaf buds enclosed by a single scale, not resinous; stamens usually less than 5; trees or shrubs with usually erect catkins .... **Salix**

POPULUS Cottonwood; Aspen; Poplar

Resinous trees with shallow root systems. Lvs alternate, simple and single toothed. Catkins drooping, appearing before the lvs. Fls subtended by an obliquely cup shaped disc; stamens 6-many. Capsule 2-4 valved.

1 Smooth bark; buds not sticky; petiole flattened when viewed in cross-section .... **P. tremuloides**
1 Rough bark; buds sticky with resins; petiole round in cross-section
  2 Leaves narrow; 4-5 times as long as they are broad .... **P. angustifolia**
  2 Leaves less than 3 times as long as they are broad
    3 Leaves as broad as they are long, leaf base round to heart-shaped; lateral buds diverging ..... **P. trichocarpa**
    3 Leaves 2 times as long as they are broad, leaf base slightly round; lateral buds appressed against the stem .... **P. balsamifera**

1. **P. angustifolia** James. Narrowleaf Cottonwood. Common along the Snake River.
2. **P. balsamifera** L. Balsam Poplar. Some taxonomists feel that this species does not occur in G.T.N.P., but occurs east and north of our area. Part of the problem is due to the relative ease with which the *Populus* species hybridize with each other.

3. **P. tremuloides** Michx. Quaking Aspen. Abundant in moist to dry habitats throughout the valley.
4. **P. trichocarpa** T. & G. Frequent 3 miles up Cascade Canyon.

SALIX Willow

Contributed by Robert D. Dorn

Shrubs or rarely trees. Buds with a single scale. Catkins usually erect to spreading. Fls subtended by 1 or 2 glands and an entire to erose, sometimes deciduous, bract. Stamens 2-8. Valves of capsule 2. *S. alba* L., a tree, is planted in Jackson and at the old Moran townsite.

1 Plants with pistillate catkins
  2 Capsules glabrous
    3 Plants 1-3 cm high, the leaves 9 mm or less long, near or above timberline .... **S. rotundifolia**
    3 Plants not as above
      4 Leaf blades mostly linear or nearly so, 6 or more times as long as wide, usually less than 12 mm wide; bracts subtending flowers yellow or greenish, deciduous
        5 Leaves green on both sides, often pubescent; bracts often lanceolate or lance-linear, pubescent or sometimes glabrate .... **S. exigua**
        5 Leaves usually glaucous or glaucescent beneath, glabrous when expanded; bracts usually broader, glabrous or sometimes pubescent at base .... **S. melanopsis**
      4 Leaf blades not linear, usually less than 6 times as long as wide, some often over 12 mm wide (rarely not expanded in fruit); bracts various
        6 Leaves glaucous or glaucescent beneath (rarely not expanded in fruit)
          7 Styles averaging 0.7 mm or less long ... **S. lutea**
          7 Styles averaging over 0.7 mm long
            8 Catkins, or some of them, at tips of twigs of previous year; styles 1-3 mm long; leaves very finely glandular-toothed .... **S. tweedyi**
            8 Catkins not at tips of twigs; styles 0.5-1.8 mm long; leaves crenate-serrate .... **S. pseudomonticola**

6 Leaves not glaucous beneath
9 Leaf blades lanceolate and long acuminate at tip when expanded; petioles with glands near base of blade on upper side ... **S. lasiandra**
9 Leaf blades sometimes lanceolate but not long acuminate; petioles usually lacking glands
10 Styles 1-3 mm long; some catkins at tips of twigs of previous year; leaves very finely glandular-toothed .... **S. tweedyi**
10 Styles 0.2-1.5 mm long; catkins not at tips of twigs; leaves entire to serrate
11 Plants mostly less than 2 m high; catkins mostly 0.8-2 cm long; stipes 0-0.8 mm long .... **S. wolfii**
11 Plants often over 2 m high; catkins mostly 2-6 cm long; stipes 0.5-4 mm long .... **S. boothii**
2 Capsules hairy
12 Plants creeping shrubs 1-8 cm high, near or above timberline
13 Leaf tip usually rounded; leaves glaucous and prominently reticulate-veined beneath; styles less than 0.5 mm long; nectaries 2 ...... **S. reticulata**
13 Leaf tip usually pointed; leaves glaucous or not beneath, usually not reticulate-veined; styles 0.3-2 mm long; nectary 1
14 Leaf blades mostly elliptic to oval, glaucous beneath, the old ones usually not persisting; catkins 1-5 cm long .... **S. arctica**
14 Leaf blades narrowly elliptic to elliptic, usually green beneath, the old ones often persisting; catkins mostly 0.6-2 cm long .. **S. cascadensis**
12 Plants erect shrubs or trees mostly over 20 cm high (*S. glauca* rarely creeping), only occasionally above timberline
15 Leaves mostly linear or nearly so, 6 or more times as long as wide, usually less than 12 mm wide; bracts subtending flowers yellow or greenish, deciduous; twigs not pruinose
16 Leaves green on both sides, often pubescent;

bracts often lanceolate or lance-linear, pubescent or sometimes glabrate . . . . **S. exigua**

16 Leaves usually glaucous or glaucescent beneath, glabrous when expanded; bracts usually broader, glabrous or sometimes pubescent at base . . . . **S. melanopsis**

15 Leaves not linear (rarely not expanded in fruit), the width various, usually less than 6 times as long as wide, or if more, the twigs usually pruinose; bracts usually not as above

17 Twigs of previous year, and sometimes those of season, pruinose, sometimes only apparent at the nodes especially behind buds

18 Catkins 15-60 mm long, densely flowered, sessile or nearly so; stipes 0.1-0.8 mm long . . . . **S. drummondiana**

18 Catkins 8-25 mm long, loosely flowered, with leafy floriferous branchlets 2-18 mm long; stipes 1-3 mm long

19 Leaves usually glaucous beneath with few hairs, some reddish; flower bracts usually blackish . . . . **S. lemmonii**

19 Leaves usually not glaucous but white-sericeous on both sides; flower bracts usually pale . . . . **S. geyeriana**

17 Twigs not pruinose (rarely so in *S. planifolia*)

20 Leaf blades narrowly elliptic, oblong, or oblanceolate, usually densely silvery hairy beneath, mostly glabrous or glabrate and green above; stipes 0.1-0.8 mm long . . . . **S. drummondiana**

20 Leaf blades often broader, not densely silvery beneath, or if so, then similar above (rarely not expanded in fruit); stipes various

21 Leaves equally green on both sides, pubescent

22 Catkins 1-5 cm long; young leaves with prominently glandular margins . . . . **S. eastwoodiae**

22 Catkins 0.8-2 cm long; young leaves

usually lacking glands on margins . . . . **S. wolfii**

21 Leaves glaucous or lighter beneath, glabrous or pubescent (rarely not expanded in fruit)

23 Stipes mostly 2-5 mm long; styles 0.4 mm or less long; twigs of year usually red-purple and appressed hairy; bark of older twigs cracked giving white-streaking appearance . . . . **S. bebbiana**

23 Stipes less than 2 mm long, or if as long as 3 mm, the styles often over 0.4 mm long and the twigs not as above

24 Plants to 1.5 m high, the catkins appearing with the leaves and on leafy floriferous branchlets 5-25 mm long; leaves and twigs often hairy . . . . **S. glauca**

24 Plants often over 1.5 m high, the catkins appearing before the leaves or sometimes with them, sessile or nearly so or rarely with mostly naked floriferous branchlets to 13 mm long; leaves and twigs sometimes glabrous to sparsely pubescent

25 Stipes 0-1 mm long; leaves, if present, elliptic; stigmas usually less than 0.5 mm long . . . . **S. planifolia**

25 Stipes 0.8-2.8 mm long; leaves, if present, mostly obovate to oblanceolate; stigmas usually over 0.5 mm long . . . . **S. scouleriana**

1 Plants without pistillate catkins, with mature leaves

26 Plants creeping shrubs less than 8 cm high, near or above timberline

27 Leaves 9 mm or less long, nearly as wide, green on both sides, the old ones persisting . . . . **S. rotundifolia**

27 Leaves mostly over 9 mm long, or if less, mostly twice or more as long as wide or else glaucous beneath, the old ones persisting or not
28 Leaf tip usually rounded, the blades glaucous and prominently reticulate-veined beneath .... **S. reticulata**
28 Leaf tip usually pointed, the blades glaucous or not, usually not prominently reticulate-veined beneath
29 Leaf blades elliptic to oval, 2-25 mm wide, glaucous beneath, the old ones usually not persisting .... **S. arctica**
29 Leaf blades narrowly elliptic to elliptic, 2-6 mm wide, usually not glaucous beneath, the old ones often persisting .... **S. cascadensis**
26 Plants mostly erect trees or shrubs over 20 cm high (rarely creeping in *S. glauca*), only occasionally above timberline
30 Leaf blades mostly linear or nearly so, 6 or more times as long as wide, usually less than 12 mm wide; twigs not pruinose
31 Leaves green on both sides, often pubescent; lower elevations .... **S. exigua**
31 Leaves usually glaucous or glaucescent beneath, glabrous; all elevations except alpine **S. melanopsis**
30 Leaf blades not linear, the width various, usually less than 6 times as long as wide, or if more, the twigs usually pruinose
32 Twigs of previous year, and sometimes those of season, pruinose, sometimes only apparent at nodes especially behind buds; leaves usually pubescent
33 Twigs of previous year mostly reddish when fresh; leaves dull green above, usually not sharply contrasting with the lower side, not leathery
34 Leaves usually glaucous beneath with few hairs, some hairs reddish .... **S. lemmonii**
34 Leaves usually not glaucous but white-sericeous on both sides .... **S. geyeriana**
33 Twigs often yellowish; leaves usually dark,

shiny green above, silvery hairy beneath, the two sides sharply contrasted, somewhat leathery .... **S. drummondiana**

32 Twigs not pruinose (rarely so in *S. planifolia*); leaves pubescent or glabrous

35 Leaf blades narrowly elliptic, oblong, or oblanceolate, usually densely silvery hairy beneath, the hairs often concealing the leaf surface or nearly so, dark green and glabrous or more sparsely hairy above .... **S. drummondiana**

35 Leaf blades often broader, not densely silvery hairy beneath, or if so, then similarly hairy above

36 Leaves glaucous or lighter beneath

37 Twigs of year usually red-purple and appressed hairy; bark of older twigs cracked giving white-streaking appearance; mature buds with depressed margins; plants often over 2 m high . **S. bebbiana**

37 Twigs and buds not as above; plants often less than 2 m high

38 Plants with mostly oblanceolate to obovate leaf blades; freshly stripped bark of living twigs of previous year usually with a "skunky" odor; shrub or tree usually over 2 m high, common on drier sites in upland forests .... **S. scouleriana**

38 Plants not as above

39 Most leaves entire or nearly so

40 Mature leaves glabrous or nearly so .... **S. planifolia**

40 Mature leaves usually obviously pubescent, rarely glabrate . **S. glauca**

39 Most leaves toothed

41 Leaves only slightly lighter beneath, very finely glandular toothed; twigs of year with long spreading hairs; mountain streambanks .... **S. tweedyi**

41 Leaves glaucous beneath, often

more coarsely toothed; twigs often glabrous or with appressed hairs; habitat various

42 Leaf blades mostly elliptic, dark green and shiny above; twigs usually red and shiny .... **S. planifolia**

42 Leaf blades mostly lanceolate to ovate or obovate, if elliptic, the leaves not shiny above and the twigs not red and shiny

43 Leaf blades usually ovate, obovate, or elliptic, the midrib and petiole often red; twigs of year pubescent ...... **S. pseudomonticola**

43 Leaf blades usually lanceolate or sometimes narrowly elliptic, the midrib and petiole usually green; twigs of year usually glabrous .... **S. lutea**

36 Leaves about equally green on both sides

44 Mature leaf blades lanceolate and long acuminate at tip; petioles with glands near base of blade .... **S. lasiandra**

44 Mature leaf blades sometimes lanceolate but not long acuminate at tip; petioles usually lacking glands

45 Leaf blades broadly elliptic, ovate, or obovate, very finely glandular toothed; twigs of year with long spreading hairs .... **S. tweedyi**

45 Leaf blades usually elliptic, lanceolate, or oblanceolate, entire or toothed, sometimes also glandular; twigs various

46 Young leaves with prominently glandular margins; mostly subalpine .... **S. eastwoodiae**

Salix branches, catkins and flowers

Salicaceae

46 Young leaves often without glandular margins; subalpine or not
47 Plants mostly less than 2 m high; mature leaves are often densely silvery pubescent .. **S. wolfii**
47 Plants often over 2 m high; mature leaves mostly sparsely pubescent to glabrous .... **S. boothii**

1. **S. arctica** Pall. Common on high peaks and in cirques.
2. **S. bebbiana** Sarg. Common in the valleys.
3. **S. boothii** Dorn. Common in the valleys.
4. **S. cascadensis** Cocker. Rare on high peaks and in cirques.
5. **S. drummondiana** Barratt ex Hook. Common in the valleys and canyons.
6. **S. eastwoodiae** Cocker. ex Heller. Subalpine.
7. **S. exigua** Nutt. Rare in valleys.
8. **S. geyeriana** Anderss. Common in the valleys.
9. **S. glauca** L. Subalpine.
10. **S. lasiandra** Benth. Common in valleys along streams. Our plants are var. **caudata** (Nutt.) Sudw.
11. **S. lemmonii** Bebb. Common in valleys.
12. **S. lutea** Nutt. Common in valleys.
13. **S. melanopsis** Nutt. Common in valleys and canyons.
14. **S. planifolia** Pursh. Rare in bogs and swamps.
15. **S. pseudomonticola** Ball. Rare in the valleys.
16. **S. reticulata** L. ssp **nivalis** (Hook.) Löve et al. Common on high peaks and in cirques.
17. **S. rotundifolia** L. ssp. **dodgeana** (Rydb.) Argus. Rare on limestone on Sheep Mt. and in Alaska Basin.
18. **S. scouleriana** Barratt ex Hook. Common in the valleys and in pine forests.
19. **S. tweedyi** (Bebb ex Rose) Ball. Uncommon along streams in canyons.
20. **S. wolfii** Bebb. Common in valleys and meadows.

SANTALACEAE Sandalwood Family

Herbs, parasitic (in ours). Lvs alt, simple, no stip. Infl cymose/corymbose. Fls uni/bisex, reg, ov inferior. Ca 3-5, Co 0, S 3-5, P 3-5*. Achene/drupe.

COMANDRA Comandra

Perennial herbs, rhizomatous, parasitic on roots of conifers and angiosperms. Lvs alt, more or less fleshy. Fls in axillary or terminal clusters; perianth greenish to white, connate; filaments hairy at the base; fruit a drupe.

1. **C. umbellata** (L.) Nutt. var. **pallida** (DC.) Jones. Bastard Toad-flax. Common in big sagebrush at such places as Windy Point, Timbered Island and Moose. A cross-section through the root will reveal a conspicuous blue dye. Plant is alternate host for a rust of *Pinus contorta* — *Cronartium comandrae.*

SAXIFRAGACEAE Saxifrage Family

Herbs/woody. Lvs usually alt, simple/compound, usually no stip. Fls usually bisex, reg, ov superior/inferior. Ca 4-5(*), Co 4-5, S 3-10, P 2-6*. Capsule/follicle/berry.

1 Leaves mostly basal, entire, oval or kidney shaped; scapes 1-flowered; flowers white; stamens 5, alternating with clusters of sterile stamens .... **Parnassia**
1 Plants lacking the above combination of features
  2 Stamens 5
    3 Petals digitate or pinnatifid .... **Mitella**
    3 Petals entire .... **Heuchera**
  2 Stamens 10
    4 Styles or style branches 3; petals lacinate or toothed; plants usually bulblet bearing .... **Lithophragma**
    4 Styles or style branches usually 2; petals entire; plants usually not bearing bulblets
      5 Styles partially connate; petals pink to deep red .... **Telesonix**
      5 Styles free above the ovary; petals white to yellow (reddish-purple in one species with opposite leaves) .... **Saxifraga**

HEUCHERA Alumroot

Perennial herbs, naked to bracteate flowering stems. Lvs basal, long petioled, palmately lobed and crenate to dentate. Fls borne in panicles to contracted spikes; calyx variable, from campanulate to saucer-shaped; petals usually 5, white to greenish yellow; stamens 5, opposite the calyx lobes; ovary half to nearly inferior; capsule many-seeded.

1. **H. parvifolia** Nutt. var. **utahensis** (Rydb.) Garrett. Infrequent on trail to Phelps Lake Lookout, also Blacktail Butte.

LITHOPHRAGMA Woodland Star

Perennial herbs with bulblet bearing rootstocks or stems. Lvs mostly basal, palmately parted or cleft, mostly reniform. Fls rather showy, borne in simple or somewhat compound racemes; petals 5, white to pink, variously lobed or entire; stamens 10, pistil with 3 carpels; ovary 1 celled; fruit a capsule.

1 Cauline leaves bulbiferous; upper stem reddish-purple .... **L. bulbifera**
1 Cauline leaves not bulbiferous; entire plant green . **L. parviflora**

1. **L. bulbifera** Rydb. Frequent in the sagebrush flats of the valley.
2. **L. parviflora** (Hook.) Nutt. Frequent in wet meadows near Two Ocean Lake.

MITELLA Mitrewort; Bishop's Cap

Perennial herbs with slender flowering stems. Lvs mostly basal, orbicular-cordate. Fls borne in simple racemes; petals 5, alternate with the sepals, greenish or white (in ours), 3 cleft, pinnatifid or entire; stamens 5 or 10; pistil with 2 carpels; fruit a capsule.

1 Stamens opposite the petals; petals greenish and pinnatifid .... **M. pentandra**
1 Stamens alternate with the petals; petals white, trilobed .... **M. stauropetala**

1. **M. pentandra** Hook. Frequent along streams in the major canyons.
2. **M. stauropetala** Piper. Infrequent along streams in the major canyons.

PARNASSIA Grass-of-Parnassus

Perennial herbs from simple rootstocks. Lvs basal, petiolate and entire. Fls single and terminal on leafless or 1 bracteate peduncles; petals 5, white; fertile stamens 5; staminodia always present. Ovary superior to half inferior; fruit a capsule.

1 Petals fringed on the lower half .... **P. fimbriata**

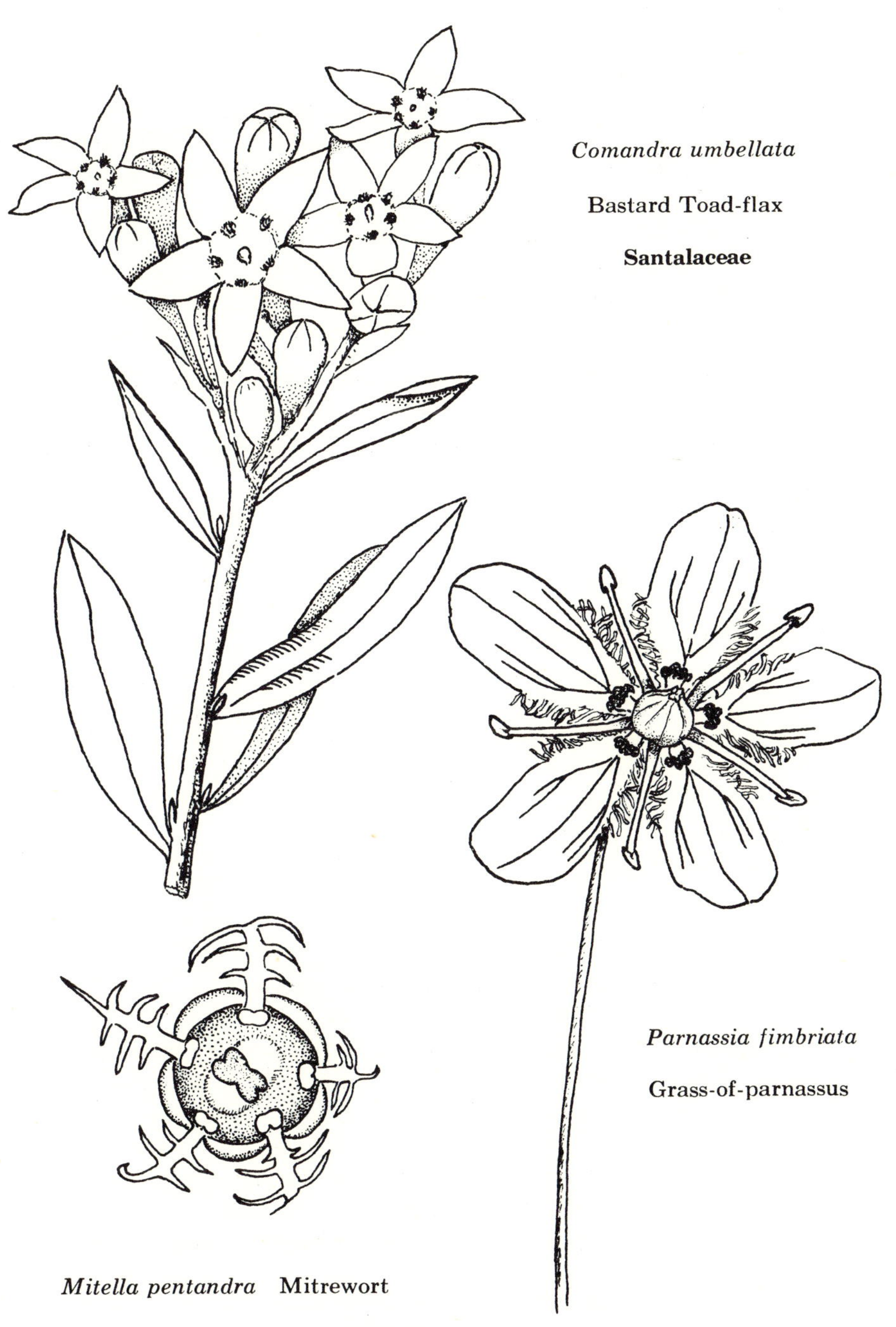

Comandra umbellata

Bastard Toad-flax

Santalaceae

Parnassia fimbriata

Grass-of-parnassus

Mitella pentandra Mitrewort

Saxifragaceae

1 Petals not fringed on the lower half
  2 Flowering stems usually bractless; petals 1 to 3-veined .... **P. kotzebuei**
  2 Flowering stems usually with a bract above the level of the basal leaves; petals 5 to 13-veined
    3 Leaf blades lanceolate to ovate often cordate; bract often clasping; petals 7-13 veined .... **P. palustris**
    3 Leaf blades elliptic to elliptic ovate; bract never clasping; petals usually 5-veined .... **P. parviflora**

1. **P. fimbriata** Konig. Frequent around the shores of Jenny Lake and along streams in the major canyons.
2. **P. kotzebuei** Cham. var. **kotzebuei.** Holly Lake Cirque, east bank.
3. **P. palustris** L. var. **montanensis** (Fern & Rydb.) C. L. Hitchc. Holly Lake Cirque.
4. **P. parviflora** DC. Frequent in wet areas along the Snake River.

SAXIFRAGA Saxifrage

Perennial herbs (in ours) with leafless or leafy flowering stems. Lvs mostly alternate, usually in a basal cluster. Fls generally cymose, but often in cymose panicles; petals 5, commonly white, sometimes spotted; stamens 10; ovary almost superior to partly inferior; fruit a capsule.

1 Leaves opposite, decussate, imbricate; petals purple .... **S. oppositifolia**
1 Leaves alternate (at least the cauline leaves); petals white or yellow, rarely pinkish
  2 Plants with slender stolons; petals yellow ..... **S. flagellaris**
  2 Plants not stoloniferous; petals various
    3 Leaves circular in outline, dentate margin, petioles long and slender .... **S. odontoloma**
    3 Leaves lanceolate, broader or reniform margins various, petioles short
      4 Leaves reniform, with shallow crenate lobes . **S. debilis**
      4 Leaves other than above
        5 Petals white with tiny purple or orange spots .... **S. bronchialis**
        5 Petals white, cream or greenish, but not spotted
          6 Stamens inserted at the edge of a narrow

bandlike gland surrounding, but not covering, the top of the ovary .... **S. occidentalis**

6 Stamens inserted at the edge of a flattened lobed disc that covers part of the ovary top

7 Flowering stems mostly 3-12 dm tall; inflorescence elongate and often compound .... **S. oregana**

7 Flowering stems 1-3 dm tall; inflorescence congested and often head-like .. .. **S. rhomboidea**

1. **S. bronchialis** L. var. **austromontana** (Wieg.) G. N. Jones. Locally common on boulders in Hidden Falls area, also around subalpine lakes of major canyons; circumboreal.
2. **S. debilis** Engelm. Rare on the ledges above Amphitheater Lake.
3. **S. flagellaris** Willd. Rare on rock ledges of the Middle Teton and Grand Teton above 12,000 ft., circumboreal.
4. **S. occidentalis** S. Wats. Frequent along small streams of major canyons.
5. **S. odontoloma** Piper. Frequent along streams of the major canyons.
6. **S. oppositifolia** L. Frequent on major peaks of the range above 9,500 ft., circumboreal.
7. **S. oregana** Howell var. **subapetala** (E. Nels.) C. L. Hitchc. Nine miles north of Jackson Lake Lodge, Hwy. 89.
8. **S. rhomboidea** Greene. Frequent in moist places — major canyons, subalpine lakes and some minor peaks such as Table Mt.

TELESONIX

Perennial herbs from short rootstocks. Lvs mostly basal, reniform, doubly crenate. Fls borne in compact panicles; petals reddish-purple; stamens 10 inserted at the top of the hypanthium; pistil with 2 carpels, ovary half inferior; styles free or partially connate.

1. **T. jamesii** (Torr.) Raf. var. **heucheriformis** (Rydb.) Bacigalupi. Frequent in moist rock crevices in canyons above 8,500 ft.

SCROPHULARIACEAE Figwort Family

Herbs (in ours). Lvs mostly opp, rarely alt or whorled,

simple. Infl various. Fls bisex, usually irreg, ov superior. Ca 5, Co 5*, S 2-4, P 2*. Capsule.

1 Corolla nearly regular; anther bearing stamens 5 . . **Verbascum**
1 Corolla usually irregular; anther bearing stamens 2 or 4
  2 Corolla with a definite spur at the base . . . . **Linaria**
  2 Corolla without a spur
    3 Anther bearing stamens 2
      4 Leaves all arising from the stem, opposite . . **Veronica**
      4 Leaves chiefly basal, the stem leaves, if present, reduced and alternate . . . . **Besseya**
    3 Anther bearing stamens 4
      5 Flowers bearing a fifth sterile stamen, often rudimentary
        6 Sterile stamen conspicuous and elongate; corolla tubular or funnelform . . . . **Penstemon**
        6 Sterile stamen rudimentary, only a scale or gland attached to the corolla; corolla less than 11 mm long
          7 Plants annual; corolla gibbous at the base . . . . **Collinsia**
          7 Plants perennial; corolla urceolate, not gibbous . . . . **Scrophularia**
      5 Flowers bearing 4 fertile stamens and no sterile stamen
        8 Upper lip of the bilabiate corolla hooded or very narrow, slightly or not at all lobed
          9 Anther sacs similar in size and position; leaves toothed to dissected never entire, often basal as well as on the stem . . **Pedicularis**
          9 Anther sacs unequally set; leaves all attached to erect stem, entire or cleft
            10 Calyx 2-parted to the base; bracts not distinctly colored . . . . **Cordylanthus**
            10 Calyx 2 or 4 cleft; bracts mostly distinctively colored
              11 Plants annual; upper lip of the corolla only slightly or not at all longer than the lower lip; calyx usually equally 4-lobed . . . . **Orthocarpus**
              11 Plants perennial; upper lip of the corolla much longer than the lower

lip; calyx usually 2 cleft and the divisions 2-lobed .... **Castilleja**

8 Upper lip of the corolla not hooded or narrow, usually distinctly lobed

12 Leaves all basal; corolla small and inconspicuous .... **Limosella**

12 Leaves well distributed along the stem; corollas showy .... **Mimulus**

BESSEYA Kittentails

Perennial herbs with fibrous roots. Lvs basal with distinct petioles, blades crenate to toothed. Fls borne in spikes or racemes; calyx of 2-4 segments; corolla lacking (in ours); stamens 2, stigma capitate, capsule compressed and loculicidal.

1. **B. wyomingensis** (A. Nels.) Rydb. Infrequent at head of Death Canyon, in valley near Cunningham Ranch, and National Elk Refuge.

CASTILLEJA Indian Paintbrush

Perennial or rarely annual, hemiparasitic herbs. Lvs alt, entire to dissected. Fls borne in bract covered spikes; calyx 4-cleft, the lobes about equal, connate in lateral pairs; corolla long, narrow and bilabiate, the upper lip (galea) forms a beak enclosing the anthers, the lower lip reduced and with 3 lobes; stamens 4; pollen sacs unequally placed; stigma capitate, entire or 2-lobed; capsule loculicidal.

Castilleja is one of our most critical genera; many species do not have clearly defined limits, and hybridization occurs between many species.

1 Plants dwarf alpine; usually less than 1 dm high; bracts and flower parts tinged with yellow or purple .... **C. pulchella**

1 Plants not dwarf alpine; usually taller than 1 dm high; bracts and flower parts tinged with yellow, pink, red or purple

2 Galea several times longer than the short lower lip; bracts showy, red or purple, rarely yellowish (except in *C. sulphurea*)

3 Calyx very deeply slit down the lower side, only teeth or lobes on the upper side; plants 4-10 dm tall .... **C. linariaefolia**

3 Calyx about equally cleft on upper and lower side
4 Bracts predominantly yellow; plants 2-5 dm tall; subalpine areas .... **C. sulphurea**
4 Bracts predominantly red or purple, yellow in rare cases
5 Leaves and bracts in upper portion usually cleft into 3-7 linear spreading lobes; bracts bright red .... **C. chromosa**
5 Leaves usually all entire; bracts entire or lobed, but rarely deeply divided
6 Stems mostly 1-3 dm tall, usually unbranched, bracts crimson; flowers mostly 20-30 mm long .... **C. rhexifolia**
6 Stems mostly more than 3 dm tall, often branched, bracts scarlet, rarely crimson; flowers mostly 30-40 mm long ...... **C. miniata**
2 Galea short, seldom over half the length of the corolla; lower lip usually more than a third the length of the galea; bracts usually yellowish
7 Calyx about equally cleft into 4 linear or triangular lobes .... **C. pilosa**
7 Calyx cut usually less than half as deeply laterally as sagittally
8 Bracts usually as long as, or longer than, the flowers, rounded or obtuse; plants of lower elevations .... **C. cusickii**
8 Bracts mostly shorter than the flowers, at least the lower ones acute; bracts and herbage grayish .... **C. flava**

1. **C. chromosa** A. Nels. No specimen seen in G.T.N.P., but it has been collected on National Elk Refuge.
2. **C. cusickii** Greenm. Along the Snake River.
3. **C. flava** Wats. Dry soils, frequently associated with sagebrush.
4. **C. linariaefolia** Benth. Frequent in the willows and sagebrush along Snake River and Pilgrim Creek.
5. **C. miniata** Dougl. Frequent to locally common in the valley, major canyons and subalpine lakes.
6. **C. pilosa** (Wats.) Rydb. Frequent on morainal soil throughout the valley, [*C. longispica* A. Nels.].
7. **C. pulchella** Rydb. Frequent on high slopes of saddle between Middle Teton and So. Teton.

8. **C. rhexifolia** Rydb. Locally common in wet areas of the valley.
9. **C. sulphurea** Rydb. Moist meadows and slopes at medium elevations.

COLLINSIA Blue-eyed Mary

Annual herbs. Lvs opp, mostly entire. Fls 1-5 in a whorl, or solitary; calyx 5; corolla bilabiate, the upper lip 2-lobed, lower lip 3-lobed; stamens 4; stigma capitate or slightly 2-lobed; capsule dehiscing along 4 sutures.

1. **C. parviflora** Dougl. Common throughout the valley, under lodgepole, aspen and sagebrush.

CORDYLANTHUS Birdbeak

Annual herbs. Lvs alt, narrow or cut into narrow divisions. Fls in small heads or compact spikes; calyx cleft to the base into dorsal and ventral segments; corolla tubular, bilabiate, dull yellowish; stamens 4 or 2; capsule flattened and loculicidal.

1. **C. ramosus** Nutt. Locally common north end of Blacktail Butte.

LIMOSELLA Madwort

Perennial, scapose herbs. Lvs in a basal rosette, entire. Fls small, white or pinkish, nearly regular; stamens 4, nearly equal; stigma capitate; capsule septicidal.

1. **L. aquatica** L. Shallow water or on wet mud.

LINARIA Toadflax; Butter-and-eggs

Annual to perennial herbs. Lvs mostly alt, sessile. Fls borne in racemes; calyx of 5 distinct sepals; corolla yellow, blue or white, spurred at the base, bilabiate; stamens 4; capsule dehiscing near the summit by pores or clefts.

1 Leaves ovate or lance-ovate, clasping the stem . . **L. dalmatica**
1 Leaves linear, not clasping . . . . **L. vulgaris**

†1. **L. dalmatica** (L.) Mill. Roadsides and other disturbed sites.
†2. **L. vulgaris** Hill. Locally common in waste places throughout the valley.

MIMULUS Monkey Flower

Annual or perennial herbs. Lvs opp, entire or toothed. Fls

axillary; calyx tubular or campanulate, 5 angled and usually 5-lobed or toothed; corolla slightly to strongly bilabiate, yellow, purple or red; stamens 4, in pairs; stigma more or less 2-lobed; capsule loculicidal.

1 Corolla pink to red or purple marked with yellow, erect plants of the mountains
  2 Plants annual, without stolons or rhizomes ..... **M. breweri**
  2 Plants perennial, with stolons or rhizomes .... **M. lewisii**
1 Corolla yellow, sometimes marked with other colors, plants of diverse habitats
  3 Upper calyx tooth definitely larger than the others; corolla strongly bilabiate .... **M. guttatus**
  3 Upper calyx tooth not conspicuously larger than the others; corolla various
    4 Plants perennial from rhizomes; corolla 1.5-3 cm long .... **M. moschatus**
    4 Plants annual, no rhizomes; corolla less than 1.5 cm long
      5 Leaves abruptly constricted to the petiole; calyx teeth acute .... **M. floribundus**
      5 Leaves tapered to the sessile base; calyx short with rounded-mucronate teeth .... **M. suksdorfii**

1. **M. breweri** (Greene) Rydb. Frequent in granitic rocks, No. Teton Canyon and Cascade Canyon.
2. **M. floribundus** Dougl. Rare in Cascade and Death Canyons up to 8,000 ft.
3. **M. guttatus DC. Locally** common in wet places and stream banks of the valley.
4. **M. lewisii** Pursh. Locally common along streams in the canyons above 8,000 ft.
5. **M. moschatus** Dougl. Frequent in the mouths of the major canyons and on the shore of Swan Lake.
6. **M. suksdorfii** Gray. Frequent in Waterfalls Canyon and along the Snake River.

ORTHOCARPUS Owl's Clover

Slender annual herbs. Lvs alt, entire to dissected. Fls in spike-like racemes, with leaf-like or colored bracts; calyx 4-cleft; corolla tubular, narrow, bilabiate, the upper lip beak-like and enclosing the anthers, the lower lip more or less puffed out into

1 or 3 inflated sacks, yellow (in ours); stamens 4; capsule loculicidal.

1. **O. luteus** Nutt. Common along the Snake River and in the sagebrush flats.

PEDICULARIS Lousewort

Perennial herbs (in ours). Lvt alt, crenate to pinnately dissected. Fls in terminal spikes or racemes; calyx irregular, 2-5 lobed; corolla yellow, white or purple, bilabiate, the upper lip arched to form a distinct beak, the lower lip generally 3-lobed; stamens 4; capsule flattened and asymmetrical.

1 Calyx lobes 2; leaves only serrate .... **P. racemosa**
1 Calyx lobes 5; leaves pinnately lobed or incised
  2 Flowers yellow .... **P. bracteosa**
  2 Flowers purple, white or pink-purple
    3 Galea (upper lip of bilabiate corolla) lacking any projection or beak .... **P. pulchella**
    3 Galea with a projection or beak
      4 Beak straight; plants 0.5-3 dm tall .... **P. parryi**
      4 Beak strongly curved; plants 1.5-7 dm tall
        5 Beak strongly upcurved like an upraised elephant's trunk; flowers purple ..... **P. groenlandica**
        5 Beak with a downward curve; flowers white or yellowish white .... **P. contorta**

1. **P. bracteosa** Benth. var. **paysoniana** (Pennell) Cronq. Frequent in the coniferous forest from 6,700 ft. up to 9,000 ft.
2. **P. contorta** Benth. Locally frequent in subalpine areas — Stewart Draw and Indian Paintbrush Canyon.
3. **P. groenlandica** Retz. Elephant's head. Common in wet meadows of the north end of the valley and extending up to some subalpine lakes.
4. **P. parryi** Gray var. **purpurea** Parry. Frequent in alpine tundra up to 12,000 ft.
5. **P. pulchella** Pennell. Rare, Sheep Mt.
6. **P. racemosa** Dougl. Frequent on the moraines of the valley and extending into the canyons to 9,000 ft.

PENSTEMON Beardtongue; Penstemon

Perennial herbs or shrubs. Lvs opp, entire or toothed, the upper ones sessile and often clasping. Fls often borne in whorls;

calyx equally 5-parted; corolla tubular, bilabiate, typically blue to purple, varying to pink or white (in ours); stamens 4, paired, fifth stamen represented by a sterile filament attached to the upper side of the corolla; capsule septicidal. Many species hybridize freely under natural conditions and therefore the genus can be difficult.

1 Anthers woolly with tangled hairs (visible with the unaided eye) .... **P. montanus**
1 Anthers glabrous or inconspicuously covered with straight hairs
  2 Corolla white; leaves usually serrate; plants more or less shrubby at the base .... **P. deustus**
  2 Corolla other than white; leaves usually entire; plants scarcely shrubby at the base
    3 Mature flowers mostly over 25 mm long
      4 Anther sacs glabrous; corollas purple to creme .... **P. whippleanus**
      4 Anther sacs with stiff rigid hairs; corollas blue to lavender
        5 Calyx lobes with erose, scarious margins **P. cyaneus**
        5 Calyx lobes acuminate with entire margins .. .. **P. cyananthus**
    3 Mature flowers mostly less than 25 mm long
      6 Leaves densely hairy on both sides ...... **P. radicosus**
      6 Leaves glabrous or sparingly hairy
        7 Inflorescence not at all glandular
          8 Anther sacs less than 1 mm long .. **P. procerus**
          8 Anther sacs 1.3-2.3 mm long ..... **P. cyananthus**
        7 Inflorescence glandular pubescent
          9 Staminode prominently exserted; calyx-lobes mostly more than 7 mm long .... **P. whippleanus**
          9 Staminode included; calyx-lobes mostly less than 7 mm long
            10 Leaves below the inflorescence glabrous, not at all ash gray .... **P. attenuatus**
            10 Leaves (all, or at least those below the inflorescence) ash gray with very short hairs .... **P. humilis**

1. P. attenuatus Dougl. ssp. **pseudoprocerus** (Rydb.) Keck. Frequent in major canyons and around subalpine lakes.

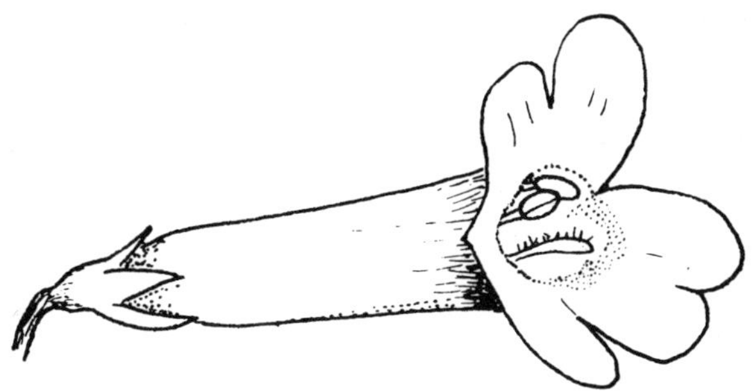

Penstemon flower

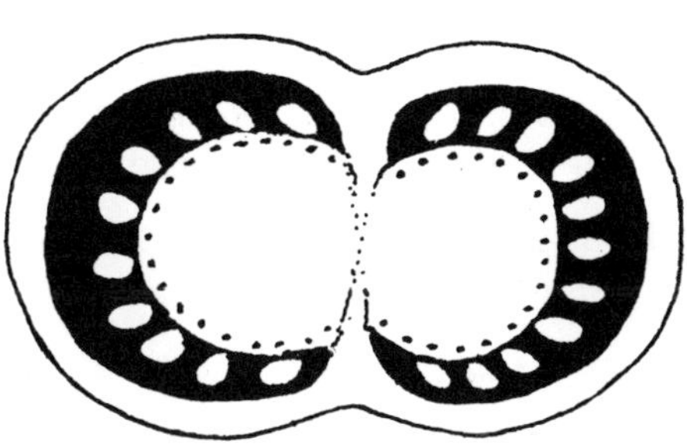

X section of fruit

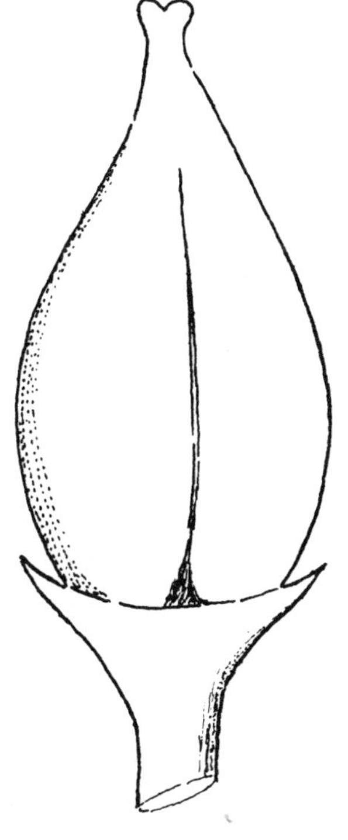

fruit

Scrophulariaceae

2. **P. cyananthus** Hook. ssp. **subglaber** (Gray) Penn. Frequent in disturbed areas throughout the valley.
3. **P. cyaneus** Penn. Infrequent in disturbed site near Swan Lake.
4. **P. deustus** Dougl. Infrequent on switchbacks above Hidden Falls, Hot Spring on Jackson Lake and mouth of Death Canyon.
5. **P. humilis** Nutt. River bottoms in the south end of the valley.
6. **P. montanus** Greene. Frequent on talus slopes above 9,500 ft.
7. **P. procerus** Dougl. Frequent in disturbed places throughout the valley.
8. **P. radicosus** A. Nels. Infrequent on the north end of Blacktail Butte.
9. **P. whippleanus** Gray. Frequent on trails in major canyons up to 9,500 ft.

SCROPHULARIA Figwort

Perennial herbs with square stems. Lvs opp, petiolate and toothed. Fls in elongate panicle; calyx deeply 5-cleft; corolla 9-11 mm long, bilabiate, greenish to brownish (in ours); fertile stamens 4, fifth stamen represented by sterile scale; capsule septicidal.

1. **S. lanceolata** Pursh. Locally frequent on rocky south facing slopes in major canyons.

VERBASCUM Mullein

Biennial or perennial herbs. Lvs alt and basal, entire toothed or pinnatifid. Fls borne in racemes, spikes or panicles; calyx of 5 sepals; corolla slightly irregular, 5 lobed, rotate, yellow or occasionally white; stamens 5, all anther bearing; stigma capitate; capsule septicidal.

†1. **V. thapsus** L. Infrequent in disturbed areas around Jackson Lake Lodge.

VERONICA Speedwell

Annual or perennial herbs. Lvs opp on stem, or alt within inflorescence, entire or toothed. Fls solitary, axillary racemes or spikes; calyx 4 parted; corolla rotate, 4-lobed, slightly irregular; stamens 2; stigma capitate; capsule often notched or lobed at the tip.

1 Leaves opposite throughout; flowers in axillary racemes; stem never terminating in inflorescence
  2 Leaves all with short petioles .... **V. americana**

2 Leaves on the middle and upper portions of the stem sessile .... **V. anagallis-aquatica**

1 Leaves (at least the upper bracts) usually alternate; stem terminating in an inflorescence

3 Plants annual; fibrous rooted .... **V. peregrina**

3 Plants perennial from underground rhizomes

4 Capsule wider than high; stem finely and closely puberulent .... **V. serpyllifolia**

4 Capsule higher than wide; stem sparsely to densely covered with loosely spreading hairs .... **V. wormskjoldii**

1. **V. americana** Schwein. Frequent in wet places of the valley.
2. **V. anagallis-aquatica** L. Along ditches and slow moving streams, specifically at Kelly Warm Springs.
3. **V. peregrina** L. var. **xalapensis** (HBK) Penn. Locally frequent in mud of drying ponds near Colter Bay.
4. **V. serpyllifolia** L. var. **humifusa** (Dickson) Vahl. Frequent in sedge meadow near Swan Lake.
5. **V. wormskjoldii** Roem. & Schult. [*V. alpina* L.]. Frequent on moist open slopes at moderate to high elevations in the mountains.

SOLANACEAE Potato Family

Herbs/woody. Lvs alt, simple, no stip. Infl often cymose. Fls bisex, reg/irreg, ov superior. Ca 5*, Co 5*, S 5, epipetalous and alt, P 2*. Berry/capsule.

HYOSCYAMUS Henbane

Poisonous annual or perennial herbs. Lvs irregularly lobed, cleft or pinnatifid. Fls showy, borne mostly on secund racemes or spikes; calyx campanulate or urceolate, 5-lobed; corolla purple reticulate on a greenish-yellow background, 5-lobed, slightly irregular; stamens 5; capsule circumscissile.

†1. **H. niger** L. Frequent; a recent invader to disturbed sites at Jackson Lake Lodge.

URTICACEAE Nettle Family

Herbs, often with stinging hairs. Lvs alt/opp, simple, usually stip. Infl various. Fls unisex, reg, ov superior/inferior. Per 0-5(*), S 4, P 1. Achene.

URTICA Nettle

Perennial herbs. Lvs opp, dentate; stinging hairs. Fls unisexual; plant monoecious or dioecious; inflorescence in axillary panicles or spikes.

1. **U. dioica** L. ssp. **gracilis** (Ait.) Seland. Stinging Nettle. Common in moist areas from the Snake River up into some of the canyons of the range.

VALERIANACEAE Valerian Family

Herbs, frequently odoriferous. Lvs opp, simple/ dissected, no stip. Infl corymbose/paniculate/cymose. Fls bisex, irreg, ov inferior. Ca 0 or reduced, Co 5*, S 1-4, epipetalous and alt, P 3*. Achene-like.

VALERIANA Valerian

Perennial herbs. Lvs opp, entire to pinnatifid, basal or cauline. Fls borne in compound cymes; calyx initially rolled up and inconspicuous, later spreading; corolla 5-lobed whitish; stamens 3; fruit an achene.

1 Plants with a stout taproot and a branched caudex; inflorescence more or less panicle-like; basal leaves tapering gradually to the petiolar base .... **V. edulis**
1 Plants with numerous, fibrous roots from a rhizome or a caudex; inflorescence like a corymb at anthesis; lower leaves mostly differentiated into blade and petiole
  2 Corolla mostly 2-4 mm long, the lobes usually shorter than the tube; some flowers unisexual ...... **V. occidentalis**
  2 Corolla mostly 4-18 mm long, the lobes scarcely half as long as the tube; flowers usually all perfect .. **V. acutiloba**

1. **V. acutiloba** Rydb. Frequent on rocky slopes at high elevations in the mountains.
2. **V. edulis** Nutt. Frequent in moist meadows and willow flats of the valley.
3. **V. occidentalis** Heller. Infrequent in open or shaded places, Phelps Lake Lookout and Skyline Trail.

VERBENACEAE Verbena Family

Herbs (in ours). Lvs opp. Infl various. Fls bisex, usually irregular, ov superior. Ca 5*, Co5*, S 2-4, P 2-5. Fruit (in ours) 4 nutlets. Plants not aromatic.

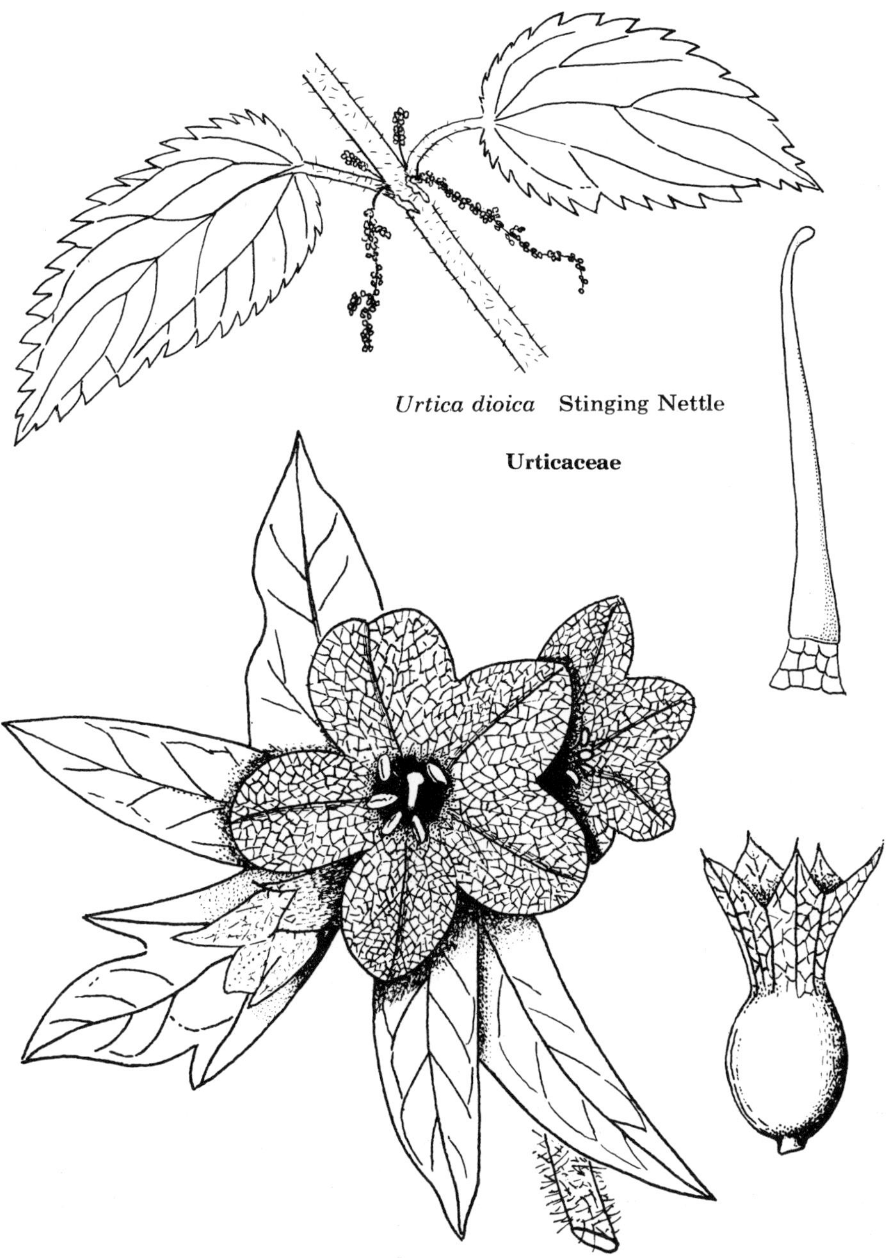

Urtica dioica Stinging Nettle

Urticaceae

Hyoscymus niger Henbane

Solanaceae

VERBENA Verbena; Vervain

Characteristics of the family.

1. **V. bracteata** Lag. & Rodr. A widespread American weed rarely seen in the valley.

VIOLACEAE Violet Family

Herbs. Lvs usually alt, simple/divided, stip. Infl various. Fls bisex, usually irreg, ov superior. Ca 5, Co 5, S 5, P 3*. Capsule/berry.

VIOLA Violet

Perennial herbs (in ours), acaulescent to caulescent. Lvs simple to compound, with stipules. Fls mostly irregular and conspicuous, but sometimes cleistogamous and inconspicuous; sepals persistent; petals blue, violet, yellow or white, the lowermost largest and spurred; stamens appearing to be fused around the pistil; ovary 1-loculed with 3 parietal points of placentation; capsule explosively dehiscent. A difficult genus with considerable hybridization between species.

1 Petals generally white, often with bluish or purplish shading
  2 Plants with annual, leafy, flower-bearing stems; petals often shaded with some yellow at the base .... **V. canadensis**
  2 Plants without annual flower stems; pedunculate flowers attached to main rhizome; petals lacking yellow
    3 Petals usually tinged with blue or violet on back; leaves 2.5-3.5 cm broad .... **V. palustris**
    3 Petals pure white except for purplish guide lines; leaves often less than 2.5 cm broad
      4 Stolons usually present; leaves usually less than 3 cm broad, glabrous on lower surface .. **V. macloskeyi**
      4 Stolons lacking; leaves usually more than 3 cm broad, often pilose on lower surface ...... **V. renifolia**
1 Petals generally blue, violet, or yellow, never white
  5 Petals bluish to purple, not yellow .... **V. adunca**
  5 Petals partially or wholly yellow
    6 Leaves reniform to cordate, usually as long as broad .... **V. orbiculata**
    6 Leaves not distinctly reniform or cordate, usually much longer than broad

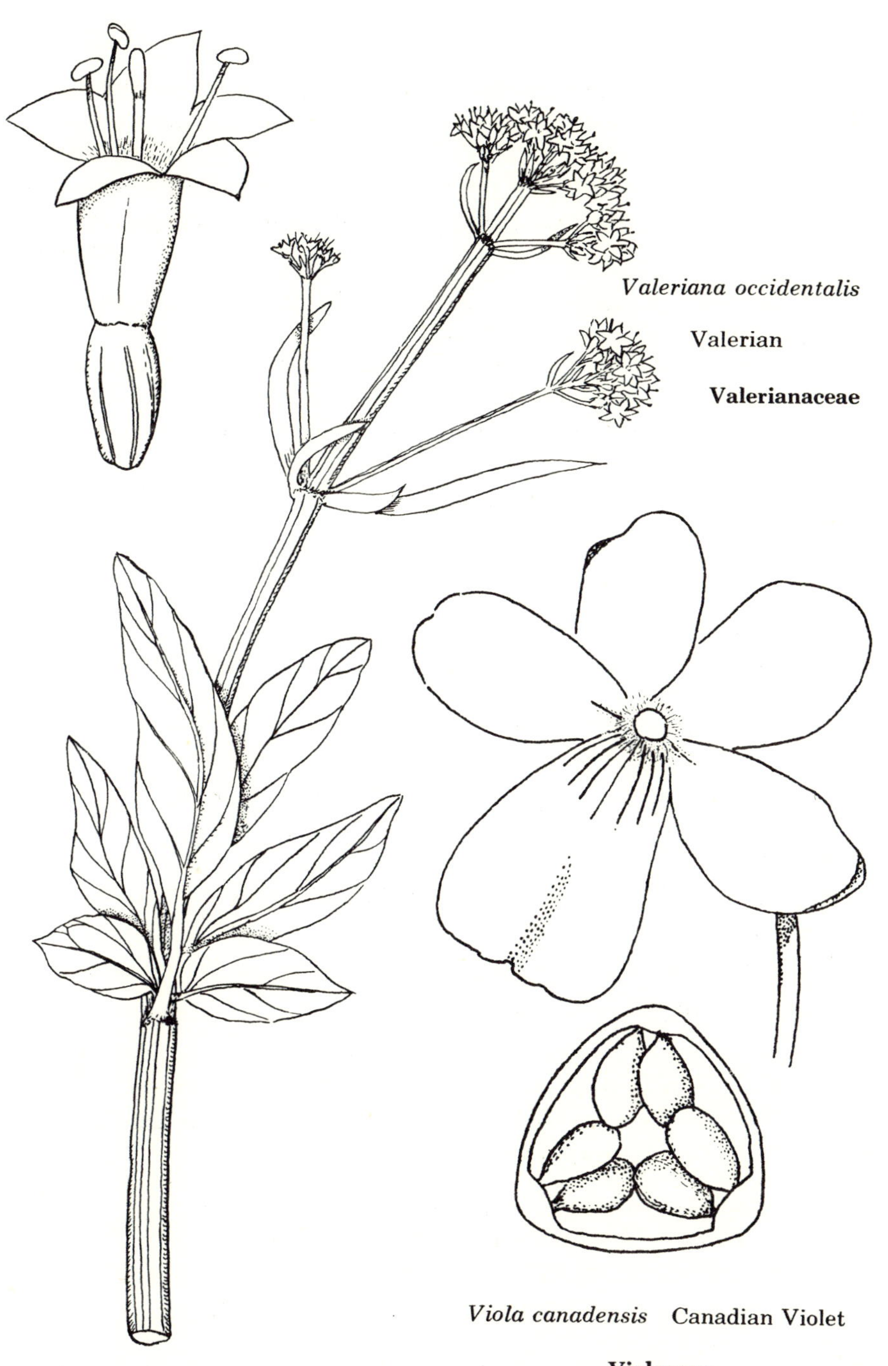

Valeriana occidentalis

Valerian

Valerianaceae

Viola canadensis Canadian Violet

Violaceae

7 Leaf blades prominently veined, usually less than 4 cm long; upper petals deep purple on back .... **V. purpurea**

7 Leaf blades not prominently veined, usually more than 4 cm long; upper petals often not purple on back .... **V. nuttallii**

1. **V. adunca** J. E. Smith. Frequent in dry to moist meadows from valley floor to subalpine lakes.
2. **V. canadensis** L. var. **rugulosa** (Greene) Hitchc. Infrequent in moist woods of the valley.
3. **V. macloskeyi** Lloyd. Infrequent in moist areas on moraines of the valley.
4. **V. nuttallii** Pursh var. **praemorsa** (Dougl.) Wats. Frequent in sagebrush or open woods of the valley.
5. **V. orbiculata** Geyer. On the north boundary, Hwy. 89.
6. **V. palustris** L. Moist meadows and streambanks; south side of Leigh Lake.
7. **V. purpurea** Kellogg. Frequent in dryish area on the east shore of Jenny Lake.
8. **V. renifolia** Gray. Frequent in Teton Canyon.

CLASS LILIOPSIDA (MONOCOTS)

ALISMATACEAE Water Plantain Family

Aquatic herbs. Lvs erect to floating, with sheathing bases. Infl racemose/paniculate. Fls uni/bisex, reg, ov superior. Ca 3, Co 3, S 6-∞, P X-∞. Achene.

SAGITTARIA Arrowhead

Marsh to aquatic herbs, often producing tubers. Lvs with sheathing bases, long petioles, arrow-shaped. Fls borne in whorls of 3; petals white, longer than the sepals; stamens usually numerous; fruit a winged achene.

1. **S. cuneata** Sheld. Frequent on the southeast shore of Two Ocean Lake, also in ponds around Jackson Lake.

CYPERACEAE Sedge Family

Grass-like herbs. Stems usually solid and often triangular. Lvs 3-ranked with closed sheaths. Infl spike-like. Fls subtended by membranous bracts, bisex/ uni. Per of scales/bristles/hairs/ absent, S 3, P 2-3*. Achene.

1 Flowers resembling a soft, feathery ball, the perianth composed of long white silky hairs .... **Eriophorum**
1 Flower clusters variable but not as above
  2 Achenes surrounded or enveloped by a small bract (perigynium) and subtended by a scale
    3 Perigynium closed except at the apex where the style protrudes .... **Carex**
    3 Perigynium split down the middle, with unsealed margins .... **Kobresia**
  2 Achenes not surrounded by a perigynium
    4 Spikelets solitary and terminal, without subtending bracts .... **Eleocharis**
    4 Spikelets one to many, subtended by one to several leafy to small scale-like bracts .... **Scirpus**

CAREX Sedge

Perennial grass-like herbs, triangular to round, solid stems. Lvs 3-ranked, closed sheaths. Plants monoecious or rarely dioecious; flowers naked, borne in spikes, each flower subtended by a small scarious bract; stamens usually 3; ovary enclosed in a sac-like bract called a perigynium; stigmas 2 or 3 protruding through the mouth of the perigynium; achenes usually 2 or 3 sided.

Small technical differences makes the identification of species in this genus frustrating, and immature specimens are usually not identifiable. A hand lens or a dissecting scope is essential.

1 Spike solitary, at the end of the stem; perigynium may contain a vestige of a rachilla in addition to the ovary .. KEY A
1 Spikes more than 1; perigynium never containing a vestige
  2 Stigmas 2 or 3; achenes lenticular or 3 angled; at least some spikes elongate and pedunculate; perigynium generally without a dorsal suture
    3 Style continuous with the achene and persistent; stigmas 3 and achene 3 angled .... KEY B
    3 Style deciduous with the other characters varying
      4 Stigmas 2; achenes lenticular .... KEY C
      4 Stigmas 3 (rarely 4); achenes 3 angled
        5 Perigynium glabrous .... KEY D
        5 Perigynium pubescent .... KEY E

2 Stigmas 2; achenes lenticular; spikes sessile and relatively short; perigynium generally with a dorsal suture
6 Spikes androgynous, or plant more or less dioecious .... KEY F
6 Spikes gynaecandrous, or some of the lateral spikes totally female
7 Perigynium planoconvex, sometimes with raised margins .... KEY G
7 Perigynium flattened or planoconvex with a thin-edged or wing-margin .... KEY H

Key A

1 Perigynia 1-3, beakless or nearly so; achenes generally 4-5 mm .... **C. geyeri**
1 Perigynia various, usually with an obvious beak; achenes generally less than 3.5 mm
2 Rachilla well developed; leaves not more than 1.5 mm wide
3 Perigynium mostly glabrous; plants generally above timberline .... **C. elynoides**
3 Perigynium short hairy at least above the middle; plants often with sagebrush .... **C. filifolia**
2 Rachilla lacking; leaves of various widths
4 Scales of female flowers becoming deciduous as the perigynium approaches maturity; perigynium beaked
5 Plants rhizomatous; moist to wet places .. **C. nigricans**
5 Plants densely caespitose, without rhizomes; meadows, ledges and rock crevices ...... **C. pyrenaica**
4 Scales of female flowers generally long-persistent; perigynium rounded or emarginate .... **C. leptalea**

Key B

1 Perigynia strongly spreading at maturity, more or less abruptly contracted to the conspicuous beak ...... **C. rostrata**
1 Perigynia ascending, tapering gradually to the poorly defined beak .... **C. vesicaria**

Key C

1 Perigynia not compressed, rounded and beakless, generally turning golden or yellow-brown; bract subtending

lowest spike with a well developed basal sheath ...... **C. aurea**

1 Perigynia various, but not as above; bract subtending lowest spike without a sheath
  2 Perigynium without nerves or nearly so on both faces .... **C. aquatilis**
  2 Perigynium prominently nerved on both faces
    3 Plants densely tufted, long rhizomes lacking; perigynium beak entire .... **C. lenticularis**
    3 Plants barely tufted, rhizomatous; perigynium beak distinctly bidentate .... **C. nebrascensis**

Key D

1 One or more bracts with conspicuous blade longer than inflorescence .... **C. oederi**
1 None of the bracts much longer than the inflorescence
  2 Stems without rhizomes .... **C. atrata**
  2 Stems with stout to creeping rhizomes
    3 Perigynium strongly inflated; spikes all erect or closely ascending .... **C. raynoldsii**
    3 Perigynium strongly flattened; lateral spikes tending to nod .... **C. paysonis**

Key E

1 Pistillate spikes 1-4 flowered; perigynium constricted above and below achene .... **C. rossii**
1 Pistillate spikes 10-20 flowered; perigynium not obviously constricted around achene
  2 Perigynium thin-walled, more or less strongly compressed, only sparsely hairy .... **C. luzulina**
  2 Perigynium thick-walled, not strongly compressed, densely covered with hairs (like velvet) ...... **C. lanuginosa**

Key F

1 Stems arising singly or in a loose cluster; plants with rhizomes
  2 Spikes small and few-flowered (1-3 pergynia per spike), well separated from each other .... **C. disperma**
  2 Spikes larger and with more flowers than above, more or less closely aggregated
    3 Spikes generally unisexual and plants dioecious; perigynia 3.5-4.6 mm long .... **C. douglasii**

3 Spikes mostly androgynous, rarely plants dioecious; perigynia generally in the range of 2.6-3.5 mm long
4 Rhizomes slender, brownish, plants 0.5-2 dm tall .... **C. stenophylla**
4 Rhizomes coarse, black or brownish black; plants generally 3-7 dm tall .... **C. praegracilis**
1 Stems arising in dense clusters, usually with short or no creeping rhizomes
5 Perigynia broadest near the base, mostly lance-triangular, gradually tapering into the beak
6 Ventral side of the leaf sheath with red dots, becoming pinkish toward the mouth .... **C. cusickii**
6 Ventral side of the leaf sheath pale or partly green, not pinkish at the mouth .... **C. neurophora**
5 Perigynia variously shaped, but seldom lance-triangular, broadest generally above the base, more or less abruptly tapered into the beak
7 Spikes few, usually less than 10; leaf sheath not red-dotted
8 Perigynium strongly planoconvex (extreme dorsal bulging of the mature perigynium), not obviously serrulate-margined; beak of perigynium shallowly bidentate .... **C. vallicola**
8 Perigynium only moderately planoconvex, more or less clearly serrulate-margined; beak of perigynium more or less bidentate .... **C. hoodii**
7 Spikes numerous more than 10, or leaf sheaths red dotted on ventral surface, or both
9 Leaves narrow, 1-2.5 mm wide; leaf sheath somewhat red dotted but not coppery-tinged ... **C. diandra**
9 Leaves wider, 3-5 mm wide; leaf sheaths red dotted and coppery-tinged .... **C. cusickii**

Key G

1 Spikes small and few flowered, usually with only 5-10 perigynia; ventral surface of perigynium nerveless or weakly nerved .... **C. laeviculmis**
1 Spikes larger and many flowered, usually with 15-25 perigynia; ventral surface of perigynium obviously nerved .... **C. canescens**

Key H

1 Bracts (at least the lower ones) much longer than the inflorescence .... **C. athrostachya**
1 Bracts mostly short and inconspicuous or the lowermost bract sometimes a little longer than the inflorescence
  2 Perigynium small, usually 2.5-3.4 mm, planoconvex
    3 Inflorescence small and compact, 8-15 mm long; plants 1.5-4 dm tall .... **C. limnophila**
    3 Inflorescence larger and looser, 15-30 mm; plants up to 7 dm tall .... **C. subfusca**
  2 Perigynium larger, usually 3.2-7.9 mm, planoconvex or flattened
    4 Perigynium more or less evidently multinerved dorsally (10 or more nerves) .... **C. phaeocephala**
    4 Perigynium otherwise, usually with fewer nerves
      5 Perigynium more or less obviously planoconvex, the cavity nearly filled by the mature achene .. .. **C. petasata**
      5 Perigynium more or less strongly flattened except where swollen by the diminuitive achene
        6 Plants generally 2-6 dm tall; perigynium 3.2-5 mm long; from the valley to near timberline .... **C. microptera**
        6 Plants generally 1-3.5 dm tall; perigynium 5-6.2 mm long; near or above timberline .... **C. haydeniana**

1. **C. aquatilis** Wahl. Circumboreal, common in the sedge meadows around the base of Signal Mt.
2. **C. athrostachya** Olney. Common in wet places north of Signal Mt.
3. **C. atrata** L. Circumboreal, Amphitheater Lake.
4. **C. aurea** Nutt. Widespread in No. America, common along the Snake River.
5. **C. canescens** L. Circumboreal, Cascade Canyon in wet locations.
6. **C. cusickii** Mack. Moose Ponds south of Pk. Hdqts.
7. **C. diandra** Schrank. Circumboreal, infrequent Swan Lake.
8. **C. disperma** Dewey. Circumboreal, streambank, Cascade Canyon.

9. **C. douglasii** Boott. Common on disturbed soil north of Signal Mt.
10. **C. elynoides** Holm. Frequent Amphitheater Lake and areas above timberline.
11. **C. filifolia** Nutt. Dry slope above Elk Ranch Reservoir south of Uhl Hill.
12. **C. geyeri** Boott. Common in the coniferous forest of the valley and canyons up to 9,700 ft.
13. **C. haydeniana** Olney. Frequent along Spread Creek.
14. **C. hoodii** Boott. Cascade Canyon and north of Signal Mt.
15. **C. laeviculmis** Meinsh. Wet places in Cascade Canyon.
16. **C. lanuginosa** Michx. Frequent in wet places north of Signal Mt.
17. **C. lenticularis** Michx. [*C. kelloggii* Boott.]. Wet places in Cascade Canyon.
18. **C. leptalea** Wahl. Wet places, Jenny Lake and Third Creek.
19. **C. limnophila** Hermann. Wet areas north of Signal Mt.
20. **C. luzulina** Olney. [C. ablata Bailey]. Cascade Canyon.
21. **C. microptera** Mack. [*C. festivella* Mack.]. Wet places near Spread Creek.
22. **C. nebrascensis** Dewey. Common in wet areas north of Signal Mt.
23. **C. neurophora** Mack. South Fork of Cascade Canyon.
24. **C. nigricans** C. A. Mey. Shores of Amphitheater Lake and Holly Lake.
25. **C. oederi** Retz. [*C. viridula* Michx.]. Circumboreal, shore of Bear Paw Lake.
26. **C. paysonis** Clokey. [*C. tolmiei* Boott. & *C. podocarpa* R. Br.]. Major canyons above 8,000 ft.
27. **C. petasata** Dewey. Meadows north of Signal Mt.
28. **C. phaeocephala** Piper. [*C. eastwoodiana* Stacey]. Common north of Signal Mt. in meadows.
29. **C. praegracilis** Boott. Common in wet places of the valley floor.
30. **C. pyrenaica** Wahl. Webb Canyon near timberline.
31. **C. raynoldsii** Dewey. Frequent along Snake River and Death Canyon.
32. **C. rossii** Boott. [*C. brevipes* Boott.]. Common on trail below Surprise Lake.
33. **C. rostrata** Stokes. Circumboreal, common along Snake River and shores of the valley lakes.

34. **C. stenophylla** Wahl. [*C. eleocharis* Bailey]. Dry to moderately moist places east of Elk Ranch Reservoir.
35. **C. subfusca** Boott. Frequent in open forest, Colter Bay Village.
36. **C. vallicola** Dewey. Frequent in sagebrush and aspen north of Signal Mt.
37. **C. vesicaria** L. Circumboreal, frequent in wet areas north of Signal Mt.

ELEOCHARIS Spike-rush

Annual or perennial herbs with terete, angular or flattened stems. Lvs mostly basal and reduced to sheaths or sheathing scales. Fls perfect, borne singly in axils of scales; perianth composed of 0-6 bristles; stamens 3 or less; style sometimes thickened toward the base, often persistent on the achene as a small projection; achene lenticular to planoconvex or more or less 3 angled.

1 Plants annual; no rhizomes .... **E. ovata**
1 Plants perennials; rhizome present
  2 Stigmas 2; achenes lenticular .... **E. palustris**
  2 Stigmas 3; achenes more or less 3 angled
    3 Tubercle (style projection at apex of achene) confluent with the achene, not forming an obvious apical cap; plants more than 1 dm tall ..... **E. pauciflora**
    3 Tubercle forming an obvious apical cap on top of achene; plants usually less than 1 dm tall .. **E. acicularis**

1. **E. acicularis** (L.) R. & S. Common on mud near Two Ocean Lake.
2. **E. ovata** (Roth) R. & S. [*E. obtusa* Schult.]. Frequent in the meadows north of Signal Mt.
3. **E. palustris** (L.) R. & S. [*E. macrostachya* Britton.]. Common in wet areas north of Signal Mt. and on the shores of String Lake; a highly variable polyploid complex.
4. **E. pauciflora** (Lightf.) Link. Frequent in wet areas north of Signal Mt.

ERIOPHORUM Cotton-grass

Perennial herbs with terete or triangular solid stems. Lvs with closed sheath and grasslike blades. Fls perfect; perianth composed of numerous persistent bristles, forming a cottony tuft at least in fruit; stamens 3; achene unequally 3 sided. This genus is closely related to *Scirpus*.

1 Leaves flattened to well beyond the middle, then triangular near the tip; achenes blackish .... **E. viridicarinatum**
1 Leaves triangular or strongly channeled; achenes light brown .... **E. gracile**

1. **E. gracile** Koch. Buffalo River north of "Govt. Bridge".
2. **E. viridicarinatum** (Englem.) Fern. Locally frequent 1 mile below Columbine Cascade, Waterfalls Canyon.

KOBRESIA Kobresia

Grass-like perennial herbs, very similar to *Carex*. Fls unisexual, solitary in the axils of papery bracts; perigynium open, with unsealed margins, merely wrapped around the ovary; stamens 3; achene 3 angled.

1. **K. myosuroides** (Vill.) Fiori. Circumboreal; not seen in G.T.N.P., but should be looked for above timberline.

SCIRPUS Bulrush

Perennial (in ours) herbs with terete or triangular, generally solid stems. Lvs with closed sheath and grasslike blades, sometimes reduced. Fls perfect, borne singly in axils of scales; perianth composed of 2-6 bristles, sometimes extending beyond the spikelet or sometimes reduced; stamens 1-3; achene usually with a short, slender projection of the style (but no tubercle as in *Eleocharis*), lenticular or more or less 3 angled.

1 Spikelets solitary, floating stems and leaves .. **S. subterminalis**
1 Spikelets 2 or more; stems stiff and emergent above shallow water
  2 Stems sharply triangular with concave sides, easily crushed between fingers .... **S. olneyi**
  2 Stems terete, not easily crushed between fingers .. **S. acutus**

1. **S. acutus** Muhl. Common in shallow ponds in the north end of the valley.
2. **S. olneyi** Gray. Common in marsh area adjacent in Polecat Creek, J. D. R. Parkway.
3. **S. subterminalis** Torr. Rare in beaver ponds along the Snake River.

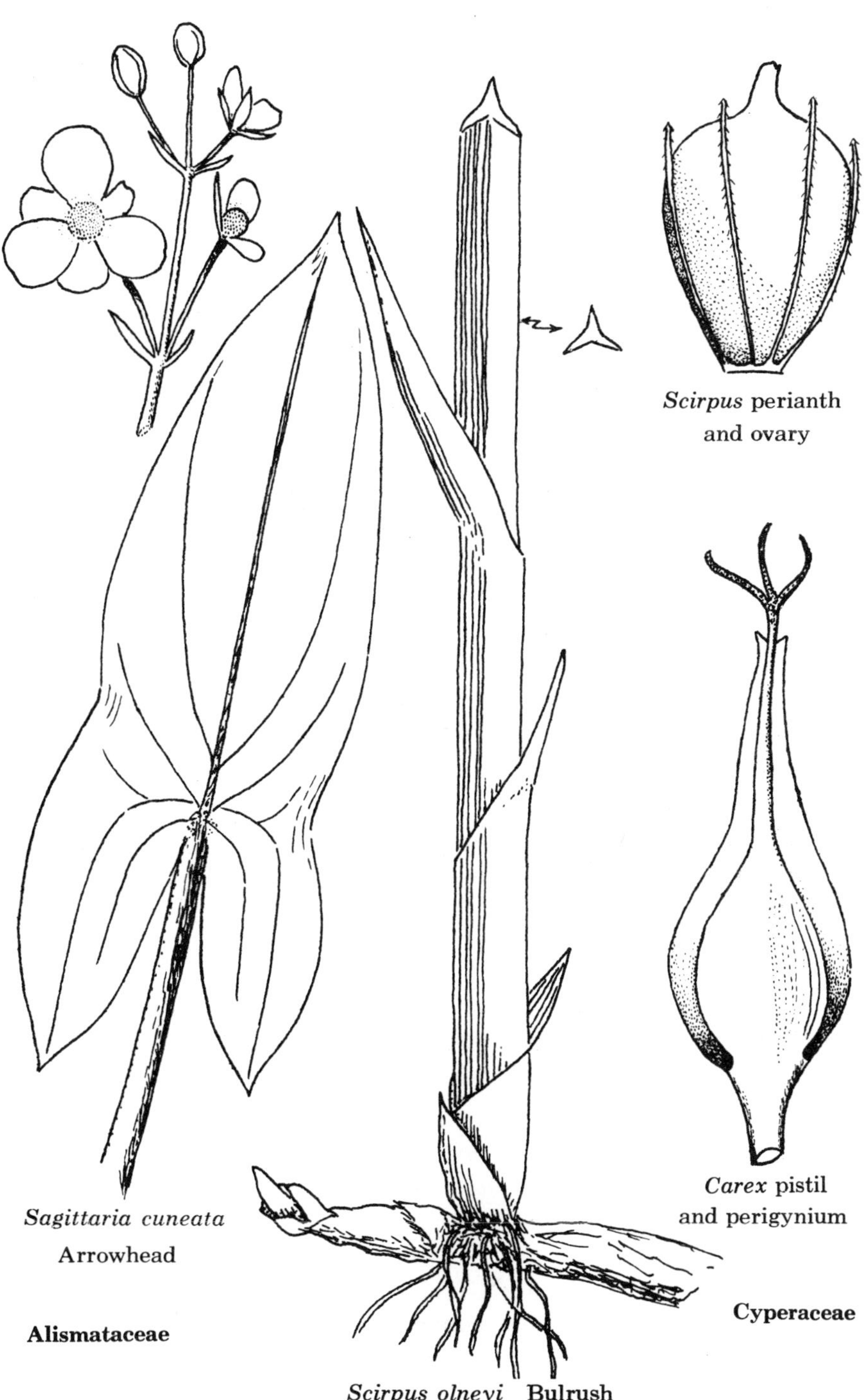

Scirpus perianth and ovary

Carex pistil and perigynium

Sagittaria cuneata Arrowhead

Alismataceae

Cyperaceae

Scirpus olneyi **Bulrush**

HYDROCHARITACEAE Frogbit Family

Aquatic herbs, submerged or floating. Lvs opp/whorled. Fls bi/unisex, reg, ov inferior. Ca 3, Co 3, S 3-∞, P 3-6*. Fruit usually berry-like.

ELODEA Waterweed; Ditchmoss

Submersed, perennial herbs, with nodally rooting stems. Lvs opp to whorled, sessile, usually denticulate margins. Fls mostly imperfect; sepals and petals 3; stamens 3-9; pistillate flowers solitary, the pedicel-like hypanthium usually elongating to bring flower to the water surface; fruit cylindric.

1. **E. canadensis** Michx. Locally frequent at Two Ocean Lake outlet and Colter Bay, Jackson Lake. Plants appear sporadically.

IRIDACEAE Iris Family

Herbs with rhizomes. Lvs narrow equitant. Infl racemose/fls solitary. Fls bisex, reg/irreg, ov inferior. Ca 3, Co 3, S 3, P 3*. Capsule.

1 Perianth segments alike and spreading; style branches slender, not at all petaloid .... **Sisyrinchium**
1 Perianth segments not similar, sepals spreading or reflexed, petals erect; style branches petaloid .... **Iris**

IRIS Iris; Flag

Rhizomatous perennials. Lvs linear, chiefly basal. Fls 1 to several in a raceme; sepals spreading and usually reflexed; petals erect; stamens opposite the sepals; capsule with 3 locules.

1. **I. missouriensis** Nutt. Not collected in the park but known to be within 10 miles of the west boundary.

SISYRINCHIUM Sisyrinchium

Perennial herbs. Lvs usually overlapping and sheathing at base, blade flattened. Fls 1-few subtended by 2 bracts; perianth segments similar, slightly connate at the base; blue to purple (in ours); filaments more or less connate; capsule with several seeds per locule.

1. **S. angustifolium** Mill. Locally frequent in moist areas along the Snake River. An extremely variable species and often regarded as several species [*S. idahoensis*, *S. samentosum*].

JUNCACEAE Rush Family

Grass-like herbs. Lvs alt/basal, sheathing. Infl cymes/ panicles/ corymbs/heads. Fls bisex, reg, ov superior. Per 6, scarious, S 3-6, P 3*. Capsule.

1 Leaf-sheaths with overlapping margins but not fused; ovules numerous; stems with spongy pith .... **Juncus**
1 Leaf-sheaths with the margins fused; ovules 3; stems hollow .... **Luzula**

JUNCUS Rush

Annual or perennial grass-like herbs, caespitose to rhizomatous, with round to flattened stems. Lvs with sheathing bases; at the point of blade diversion the sheath often forms round to pointed projections (called auricles). Fls small, borne in capitate to paniculate clusters; perianth undifferentiated, greenish to purplish brown; stamens usually 6 or 3; capsule 1-3 loculed, seeds numerous.

1 Plants annual; generally less than 1 dm tall .... **J. bufonius**
1 Plants perennial; usually more than 1 dm tall
  2 Inflorescence appearing lateral, the lowest bract of inflorescence looks like a continuation of the stem
    3 Stems usually bearing 1-4 flowers; subalpine or alpine plants
      4 Highest basal sheath without a blade or rarely 1 cm long; capsule with small terminal notch .... **J. drummondii**
      4 Highest basal sheath with a conspicuous blade, 2-7 cm long; capsule often acute .... **J. parryi**
    3 Stems usually bearing more than 7 flowers; plants often not subalpine or alpine
      5 Inflorescence small and tight, mostly 7-15 flowered; stems less than 1.6 mm thick **J. filiformis**
      5 Inflorescence often lax and loose, more than 20 flowers; stems often over 1.6 mm thick .. .. **J. balticus**

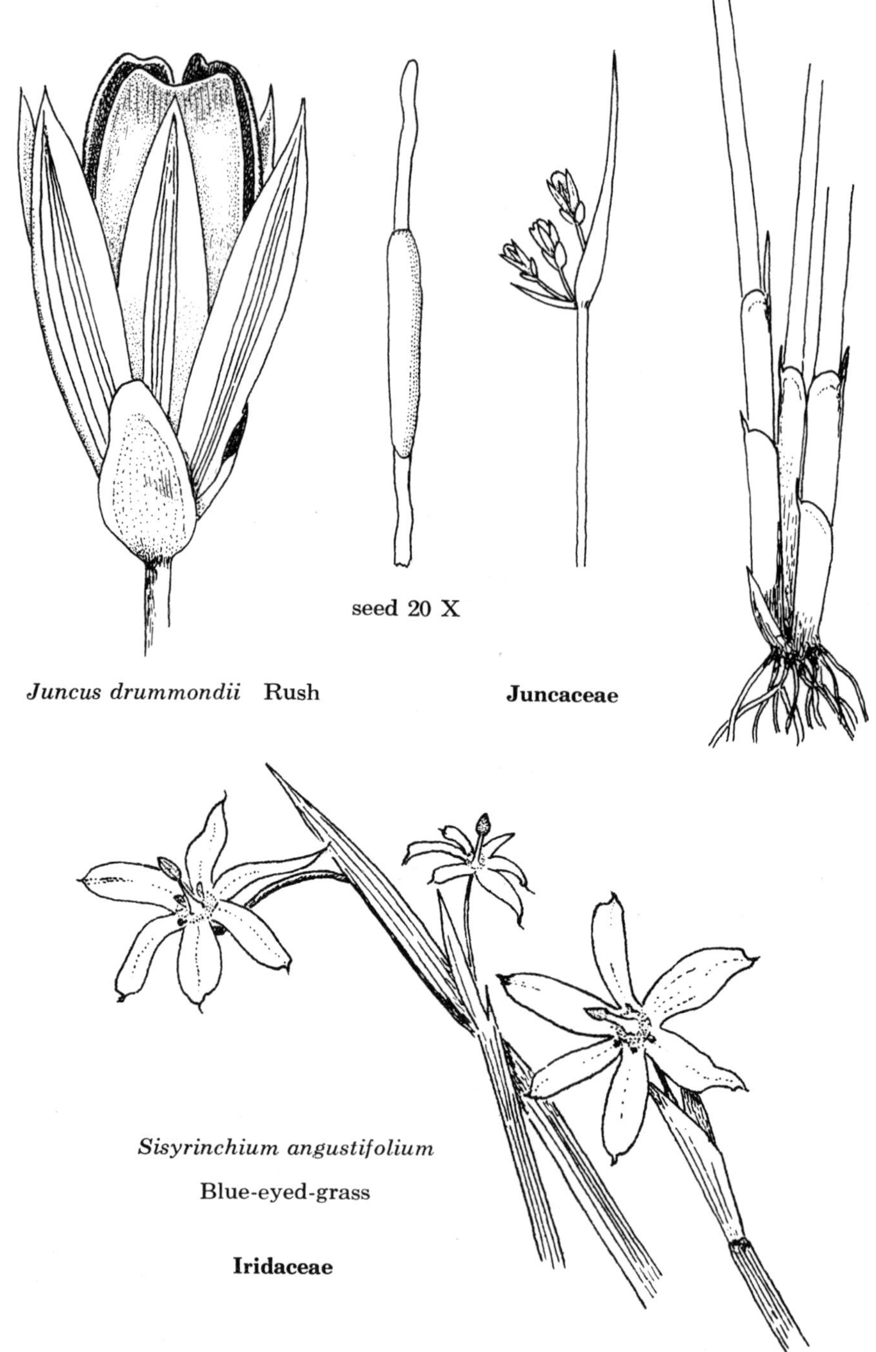

Juncus drummondii Rush

Juncaceae

Sisyrinchium angustifolium

Blue-eyed-grass

Iridaceae

2 Inflorescence terminal, involucral bract not appearing as a continuation of the stem
- 6 Leaf blades flattened laterally, equitant
  - 7 Seeds less than 0.5 mm long, with a sharp point at each end; style less than 1 mm long .... **J. ensifolius**
  - 7 Seeds between 0.8-1 mm long, with prominent appendages at each end; style about 1 mm long .... **J. tracyi**
- 6 Leaf blades not flattened laterally, not equitant
  - 8 Leaves round in cross-section, septate; plants mostly alpine **J. mertensianus**
  - 8 Leaves either dorsiventrally flattened or semi-terete, not septate; plants not alpine
    - 9 Flowers borne in heads ... **J. longistylis**
    - 9 Flowers borne singly in congested terminal inflorescence, but not in a head .... **J. confusus**

1. **J. balticus** Willd. Locally common in wet areas along the Snake River.
2. **J. bufonius** L. Common on moist sand in vicinity of the Oxbow Bend.
3. **J. confusus** Cov. Frequent in sedge meadows along the Snake River.
4. **J. drummondii** Meyer. Locally frequent around subalpine lakes.
5. **J. ensifolius** Wikst. var. **montanus** (Engelm.) Hitchc. [*J. saximontanus* A. Nels.]. Common in wet areas along the Snake River.
6. **J. filiformis** L. Frequent in sedge meadows along the Snake River.
7. **J. longistylis** Torr. Common in wet areas in vicinity of the Oxbow Bend.
8. **J. mertensianus** Bong. Frequent along stream banks and lake margins above 9,000 ft.
9. **J. parryi** Engelm. Frequent in wet meadows of the valley and subalpine slopes.
10. **J. tracyi** Rydb. Frequent in the main couloir on the east face of Symmetry Spire, 8,000 ft.

LUZULA Woodrush

Perennial, grass-like herbs. Lvs flat, fringed with hairs, with closed sheathing base. Fls borne in spike-like to umbellate or panicle-like inflorescences; perianth greenish to brown or purplish, often thin and dry; stamens 6; capsule 1-loculed, 3 seeded.

1 Flowers in 1 or more spike-like panicles .... **L. spicata**
1 Flowers at the ends of branches of an open panicle
  2 Anthers much longer than the filaments (0.8-1.5 mm); style about 1 mm; perianth dark brown or purplish brown .... **L. hitchcockii**
  2 Anthers generally shorter than the filaments; style seldom over 0.5 mm long; perianth often pale
    3 Bracts and bracteoles of inflorescence finely fringed; stem leaves mostly 2-3 mm broad .... **L. piperi**
    3 Bracts and bracteoles mostly erose and not fringed; stem leaves generally 3-10 mm broad ...... **L. parviflora**

1. **L. hitchcockii** Hamet-Ahti. [*L. glabrata,* misapplied]. Common in Holly Lake area.
2. **L. parviflora** (Ehrh.) Desv. Not seen in G.T.N.P., but reported in Y.N.P. A widespread species.
3. **L. piperi** (Cov.) Jones [*L. wahlenbergii* of authors]. Frequent, Timberline Lake and Amphitheater Lake.
4. **L. spicata** (L.) DC. Near summit of Table Mt.

JUNCAGINACEAE Arrow-Grass Family

Herbs of marshy places. Lvs terete, rush-like. Infl raceme/ spike. Fls uni/bisex, reg, ov superior. Per usually 6, S 4-6, P 4-6. Follicular fruits.

TRIGLOCHIN Arrow-grass

Rhizomatous herbs of wet places, often saline or alkaline conditions. Lvs terete. Fls perfect, 3-merous in spike-like racemes, greenish; carpels 3 or 6; fruit follicular.

1. **T. maritimum** L. Frequent west of Huckleberry Hot Springs, J. D. R. Parkway.

LEMNACEAE Duckweed Family

Small floating aquatics, flat disk-like bodies bearing 1 or more roots/rootless. Fls unisex, rare. Per 0, S 1-2, P 1. Utricle. Reproducing by budding.

1 Rootlets solitary per disk-like body .... **Lemna**
1 Rootlets several per disk-like body .... **Spirodela**

LEMNA Duckweed

Disk-like body flattened, 1-5 nerved with a single rootlet, usually in colonies. Fls generally 3 (2 staminate and 1 pistillate) in a marginal pocket of the disk; fruit 1-seeded.

1 Disks oblong to lanceolate, often with slender stalks ... . **L. trisulca**
1 Disks oval, without stalks
  2 Disks nerveless, 2-3 times as long as broad ... **L. valdiviana**
  2 Disks usually 3 nerved, generally less than twice as broad .... **L. minor**

1. **L. minor** L. Locally common in small ponds in the vicinity of Jackson Lake Dam.
2. **L. trisulca** L. Infrequent in small ponds of the valley.
3. **L. valdiviana** Phil. Locally frequent, Kelly Warm Spring.

SPIRODELA Spirodela

Disk-like bodies flattened, suborbicular, 7-12 nerved with several rootlets, usually in colonies. Fls usually 3 in a marginal pocket of the disk; ovary with 2 ovules.

1. **S. polyrhiza** (L.) Schleid. Moose Ponds, west and south of Jenny Lake.

LILIACEAE Lily Family

Herbs (in ours) from bulbs/corms/rhizomes. Lvs basal/cauline, alt/whorled. Infl various. Fls bisex, reg or nearly so, ov superior. Ca 3, Co 3, S 6, P 3*. Capsule/berry.

1 Perianth with distinct calyx and corolla .... **Calochortus**
1 Perianth in two whorls of similar segments
  2 Inflorescence a terminal umbel or head

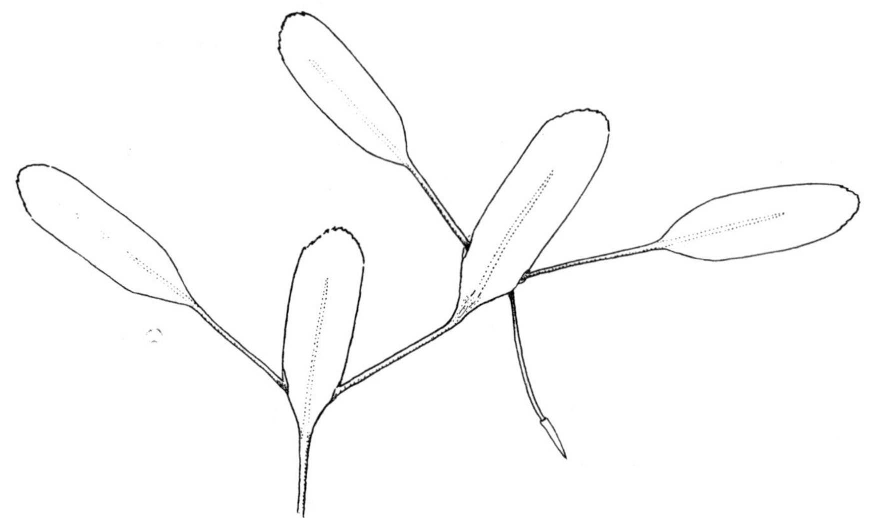

Lemna trisulca Duckweed

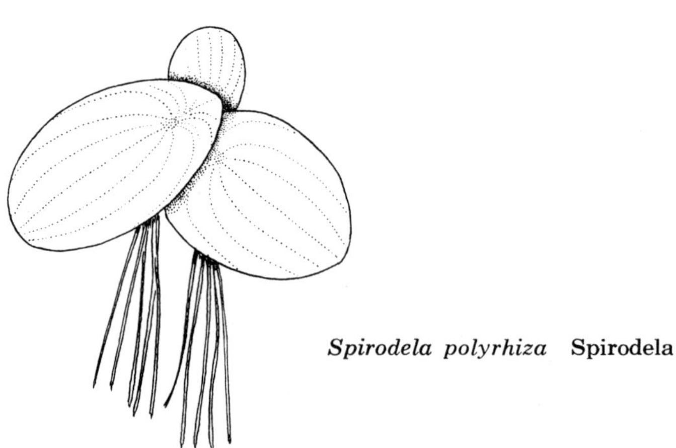

Spirodela polyrhiza Spirodela

Lemnaceae

3 Perianth segments 2-4 cm long, fused at least half the length .... **Brodiaea**
3 Perianth segments less than 2 cm long and distinct .... **Allium**
2 Inflorescence racemose to paniculate or flowers solitary
4 Leaves linear, persistent from year to year, stiff .. .. **Xerophyllum**
4 Leaves various, but not persisting from year to year
5 Leaves mostly basal, usually with a naked flowering stem
6 Styles 3, separated almost to the ovary
7 Plants with rhizomes; leaves 2-ranked . **Tofieldia**
7 Plants with a bulb; leaves not 2-ranked .. .. **Zigadenus**
6 Styles 1 or 3, but connate more than half the length
8 Leaves linear, less than 3 mm wide; perianth segments with a basal gland ...... **Lloydia**
8 Leaves various, but always more than 3 mm wide; perianth lacks glands
9 Flowers usually 10 or more; leaves several, linear .... **Camassia**
9 Flowers usually less than 5; leaves mostly 2, lanceolate or oblong .. **Erythronium**
5 Leaves usually on the flowering stem
10 Plants from a scaly bulb; leaves often whorled, linear to lanceolate .... **Fritillaria**
10 Plants with rhizomes; leaves mostly ovate, never whorled
11 Flowers 1 to 3 borne along the stem or in small clusters at the ends of the branches
12 Flowers 1 or 2 at a node on a slender peduncle .... **Streptopus**
12 Flowers 1 or 2 at the tips of branches on stout pubescent pedicels .... **Disporum**
11 Flowers 5 to many in a terminal raceme or panicle
13 Style 1; plants 2-9 dm tall; leaves without prominent veins .... **Smilacina**
13 Styles 3; plants usually more than 10 dm tall; leaves with prominent veins .. .. **Veratrum**

ALLIUM Wild Onion

Perennial herbs bearing an onion-like odor, from deep seated bulbs. Lvs 1-several, usually linear. Fls several to many in an involucrate umbel; perianth segments 6; stamens 6; ovary with 3 locules; capsule loculicidal.

1 Flowering from the center outward; leaves cylindrical and hollow .... **A. schoenoprasum**
1 Flowering from outside inward; leaves flat or chanelled, solid
  2 Bulb coats persisting as a network of fibers ...... **A. geyeri**
  2 Bulb coats never with a network of fibers ... **A. brevistylum**

1. **A. brevistylum** S. Wats. Frequent along Snake River below the Jackson Lake Dam.
2. **A. geyeri** S. Wats. var. **tenerum** M. E. Jones. Locally common ¼ mile downstream from Jackson Lake Dam. This variety has many bulbils replacing many flowers.
3. **A. schoenoprasum** L. Frequent along Snake River and into the canyons up to 8,500 ft.

BRODIAEA Brodiaea

Scapose herbs from deep scaly corms. Lvs 1 to 5, linear, long and often appearing before the scape. Fls few to many borne in an involucrate umbel; perianth segments 6, connate at least half the length; stamens 6; style 1; stigmas 3; capsule loculicidal.

1. **B. douglasii** – Wats. Infrequent in area of the Sawmill Ponds (near Moose).

CALOCHORTUS Sego Lily; Mariposa

Perennial herbs from deep bulbs. Lvs few, linear. Fls solitary to few, mostly showy, calyx and corolla distinct, the inner petals broad and colored, usually spotted, with a hairy gland near the base; stamens 6; ovary triangular, 3 locules; capsule dehiscing along septum.

1. **C. eurycarpus** Wats. Infrequent on the east side of valley on Gros Ventre Slide Road also reported on Blacktail Butte.

CAMASSIA Blue Camas

Scapose perennials from deeply buried bulbs. Lvs basal, linear and numerous. Fls borne in racemes, perianth segments

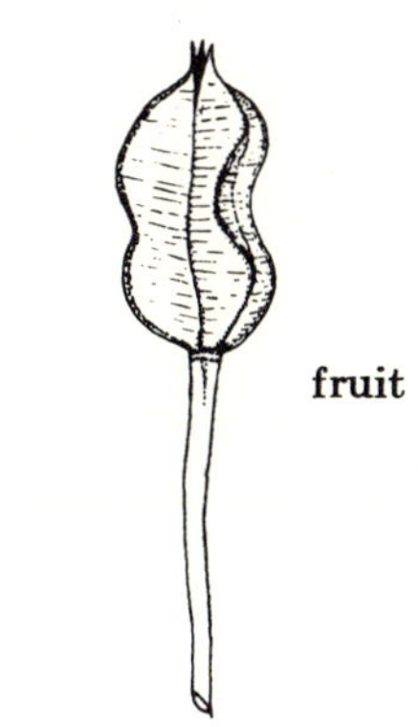

Calochortus eurycarpus
Sego Lily

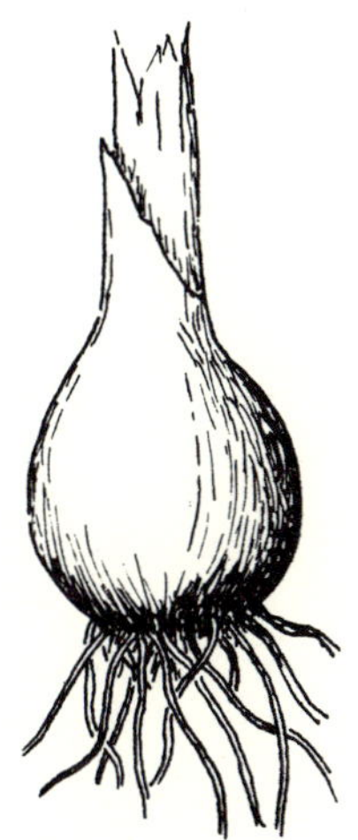

Camassia quamash Blue Camas

Liliaceae

similar and distinct, white to deep blue or violet; stigmas 3; capsule opening directly into the locules.

1. **C. quamash** (Pursh) Greene. Locally frequent in wet meadows of the valley north of Jackson Lake Dam. Bulbs are edible.

DISPORUM Fairy Bells; Wartberry

Perennial herbs from rhizomes, branched. Lvs rounded or cordate, sessile, obvious veins. Fls 1-2 on pubescent pedicels, creamy white, often concealed by the leaves; perianth segments distinct; ovary 3-loculed; fruit a yellow to red (in ours) berry.

1. **D. trachycarpum** (Wats.) Benth. & Hook. Frequent along the Snake River and in the mouth of Death Canyon.

ERYTHRONIUM Glacier Lily; Avalanche Lily; Fawn Lily; Dogtooth Violet

Perennial herbs from deep positioned corms. Lvs elliptical, in a basal pair. Fls 1-5 in a loose raceme or a naked peduncle, nodding, yellow (in ours); perianth segments separate and similar; anther basifixed; fruit dehiscing into the locules. (The color of the anthers in our species varies).

1. **E. grandiflorum** Pursh. Common in the North Fork of Cascade Canyon around melting snowbanks.

FRITILLARIA Fritillary

Perennial herbs from bulbs bearing attached bulblets; stem unbranched. Lvs alt to whorled, linear to lanceolate. Fls 1-2, usually nodding, yellow or purplish mottled (in ours); perianth segments alike, distinct; stamens shorter than the perianth; capsule loculicidal, sometimes winged.

1 Flowers yellow, turning reddish with age; only one style .... **F. pudica**
1 Flowers purplish and spotted; styles 3 .... **F. atropurpurea**

1. **F. atropurpurea** Nutt. Leopard Lily, Purple Spot Fritillary. Frequent on moraines in the coniferous forests of the valley.
2. **F. pudica** (Pursh) Spreng. Yellow Bell. Frequent in the sagebrush flats; flowering in the early spring.

LLOYDIA Alplily

Perennial herbs from short rhizomes. Lvs alt, linear, basal and surrounded by persistent bracts. Fls solitary, sometimes 2, perianth segments white with greenish or purple veins; stamens 6, anthers basifixed; style 1; loculicidal capsule.

1. **L. serotina** (L.) Sweet. Infrequent at the head of Death Canyon and gravelly ridges on Rendezvous Peak.

SMILACINA False Solomon's Seal

Perennial herbs from rhizomes, unbranched flowering stems. Lvs alt, numerous and many nerved. Fls borne in racemes or panicles; perianth segments white, distinct; stamens 6, versatile; style 1; fruit a globose, greenish to red berry.

1 Flowers usually in a simple raceme; stamens shorter than the perianth segments .... **S. stellata**
1 Flowers in a panicle; stamens exceeding the perianth segments .... **S. racemosa**

1. **S. racemosa** (L.) Desf. Frequent around the shores of the valley lakes and along the Snake River.
2. **S. stellata** (L.) Desf. Frequent in moist places of the valley.

STREPTOPUS Twisted Stalk

Perennial herbs from rhizomes, simple to branched stems. Lvs alt, ovate to oblong-lanceolate with clasping bases. Fls 1-2 on a single peduncle borne beneath the leaves, cream colored; perianth segments similar and distinct; stamens 6; fruit a yellow to red berry.

1. **S. amplexifolius** (L.) DC. var. **chalazatus** Fassett. Frequent along streambanks in many of the canyons.

TOFIELDIA Tofieldia

Perennial herbs from short rhizomes. Lvs linear, basal and more or less 2-ranked. Fls borne in racemes; perianth segment distinct, white; stamens 6, anthers basifixed; fruit a septicidal capsule.

1. **T. glutinosa** (Michx.) Pers. var. **montana** (Hitchc.) Davis. Infrequent around the shores of the valley lakes, also some subalpine lakes above 9,000 ft.

VERATRUM False Hellebore; Skunk Cabbage

Perennial herbs up to 2 m tall, from rhizomes. Lvs alt, numerous, broad and prominently veined. Fls borne in panicles;

perianth segments white to yellow or green; stamens 6, shorter than the perianth; fruit a septicidal capsule.

1. **V. californicum** Durand. Locally frequent in wet meadows in vicinity of Phelps Lake and north boundary of park on Hwy. 89.

XEROPHYLLUM Beargrass

Perennial herbs up to 15 dm tall, from a short rhizome. Lvs grass-like, persistent and wiry. Fls numerous, borne in a raceme, long pedicels; perianth segments similar and distinct, cream colored; stamens 6, anthers basifixed; fruit a loculicidal capsule; flowering stalk dies down after fruiting.

1. **X. tenax** (Pursh) Nutt. Locally frequent in open woods at the east end of Grassy Lake Dam; also in the 1974 burn area in mouth of Waterfalls Canyon.

ZIGADENUS Death Camas

Perennial herbs from deeply positioned bulbs. Lvs linear, mostly basal. Fls borne in racemes or panicles; perianth segments usually similar, white to yellow-green; stamens equal to or longer than the perianth; styles 3; fruit a capsule.

1 Perianth segments 7-11 mm long, not clawed .... **Z. elegans**
1 Perianth segments less than 7 mm long; inner segment clawed
  2 Inflorescence usually paniculate; polygamous flowers .... **Z. paniculatus**
  2 Inflorescence usually racemose; flowers bisexual .... **Z. venenosus**

1. **Z. elegans** Pursh. Infrequent on rocky slopes and shores of subalpine lakes.
2. **Z. paniculatus** (Nutt.) Wats. Frequent in sagebrush flats; Elk Island in Jackson Lake.
3. **Z. venenosus** Wats. Locally frequent in major canyons above 8,000 ft.

ORCHIDACEAE Orchid Family

Herbs, sometimes saprophytic. Lvs alt. Infl racemose/fls solitary. Fls bisex, irreg, ov inferior. Ca 3, Co 3, S 1 or 2, P. 3*. Capsule.

1 Plants without chlorophyll .... **Corallorhiza**
1 Plants with clorophyll
  2 Flowers with a distinct spur projecting down from the base of the lip .... **Habenaria**
  2 Flowers without a spur
    3 Plants with a single leaf; flowers 1 or 2 .... **Calypso**
    3 Plants with 2 or more leaves; flowers usually several to many
      4 Leaves 2, opposite about mid-length on the stem .... **Listera**
      4 Leaves usually more than 2, alternate or basal
        5 Leaves mostly basal, usually mottled; plants rarely in swampy areas .... **Goodyera**
        5 Leaves cauline, alternate, not mottled, plants usually in wet habitats .... **Spiranthes**

CALYPSO Fairy Slipper

Glabrous herb 5-20 cm tall with a more or less spherical corm. A single leaf produced at the top of the corm. Flower single, occasionally 2; sepal and petals alike, lanceolate; lip much larger than other perianth parts, pink to purple, lip variously spotted; casule erect. Only one species.

1. C. bulbosa (L.) Oakes. Infrequent in deep shade and local areas of decaying wood, in the valley.

CORALLORHIZA Coralroot

Yellowish to brownish-red perennials with coral-like rhizomes. Lvs lacking; stems bearing several sheathing bracts, no chlorophyll. Fls borne in bracteate racemes, yellow to reddish-brown or purple; lip short-clawed, simple or with lateral lobes near the base; capsule pendent. Occasional albino specimens occur displaying a pale yellow color throughout.

1 Sepals and petals pinkish with 3-5 distinct reddish-brown or purple stripes; lip margin without lobes .... **C. striata**
1 Sepals and petals yellow to pink to wine red, never striped; lip often lobed
  2 Sepals yellow or greenish-yellow, 4-6 mm long, 1 nerved .... **C. trifida**
  2 Sepals often reddish, 6-13 mm, usually 3 nerved

Corallorhiza striata Coralroot **Orchidaceae**

3 Lip non-lobed, white, usually with a few purple spots, crenulate margin, upturned .... **C. wisteriana**
3 Lip usually lobed or toothed on the side, margin generally not upturned
4 Spur lacking or only a slight bulge; lip white to cream, usually with reddish spots .... **C. maculata**
4 Spur 1-3 mm long; lip pink to reddish purple, without spots .... **C. mertensiana**

1. **C. maculata** Raf. Frequent in morainal soil around the shores of the valley lakes.
2. **C. mertensiana** Bong. Frequent in lodgepole pine forests of the valley.
3. **C. striata** Lindl. Rare; appears sporadically in the coniferous forest of the valley.
4. **C. trifida** Chat. Rare, spruce forest west shore of Leigh Lake and Jackson Lake.
5. **C. wisteriana** Conrad. Infrequent, Nature Trail Island at Colter Bay. The non-lobed lip will help to identify this species from *C. maculata.*

GOODYERA Rattlesnake Plantain

Scapose herbs with creeping rootstocks. Lvs all basal, usually white reticulate (in ours). Fls inconspicuous, borne in spike-like raceme, whitish, glandular hairy; upper sepal and lateral petals united, forming a hood over the lip, the latter more or less saccate toward the base with a straight or recurved tip; capsule ascending.

1. **G. oblongifolia** Raf. Rare at the south end of String Lake on the String Lake Trail and on second switchback above Hidden Falls Bridge.

HABENARIA Rein Orchid; Bog Orchid

Glabrous perennials, often with fleshy or tuberous roots. Lvs basal to cauline, mostly with a sheathing base. Fls small in spike-like racemes, white to yellowish-green; base of lip producing a spur.

1 Leaves on the lower third of the stem; the plant scapose; flowers greenish
2 Leaf solitary, basal; lip 5-20 mm long; plants usually in moist areas .... **H. obtusata**
2 Leaves 1-5; lip less than 6 mm long; plants mostly of dry areas .... **H. unalascensis**

1 Leaves several to many, all cauline; flowers often white
  3 Flowers white; lip rhombic-lanceolate, dilated at the base .... **H. dilatata**
  3 Flowers greenish; lip lanceolate to linear, not dilated at the base .... **H. hyperborea**

1. **H. dilatata** (Pursh) Hook. Infrequent in wet areas on moraines and the major canyons up to 8,500 ft.
2. **H. hyperborea** (L.) R. Br. Infrequent along the Snake River near Biological Station.
3. **H. obtusata** (Banks) Richards. Rare at the base of the north slope of Signal Mt. also in Webb Canyon.
4. **H. unalascensis** (Spreng.) Wats. Infrequent in the shade of lodge-pole on the eastshore of Jackson Lake.

LISTERA Twayblade

Small rhizomatous herbs. Lvs broad, sessile, and opposite about mid-length of the stem. Fls borne in bracteate racemes, greenish; sepals and petals similar; lip much longer than the sepals, shallowly or deeply bilobed (in ours).

1 Lip dissected about half the length into 2 divergent lobes; leaves subcordate .... **L. cordata**
1 Lip only shallowly bilobed; leaves not cordate
  2 Lip puberulent, at least along the margins, oblong in outline and not narrowed toward the base ...... **L. borealis**
  2 Lip often not puberulent, usually narrowed considerably near the base
    3 Lip with a fringe of marginal hairs, tip of lip distinctly notched, abruptly narrowed to a short claw .... **L. convallarioides**
    3 Lip margin glabrous, tip rounded to only slightly notched, gradually narrowed to the base ..... **L. caurina**

1. **L. borealis** Morong. Rare, often in moss along streams in the mountains.
2. **L. caurina** Piper. Rare, Webb Canyon and White Grass Ranger Station.
3. **L. convallarioides** (Sw.) Nutt. Rare on the west shore of Jenny Lake.
4. **L. cordata** (L.) R. Br. Locally frequent at the edge of pond on Elk Island, Jackson Lake, also Indian Paintbrush Canyon Trail.

SPIRANTHES Ladies' Tresses

Small herbs with fleshy, fascicled roots. Lvs basal or cauline. Fls in a dense spike, usually spirally twisted in longitudinal rows, white (in ours); the sepals and lateral petals appear to be fused forming a hood around the column; lip bends down.

1. **S. romanzoffiana** Cham. & Schl. Frequent in moist to swampy areas along the Snake River and ponds near Colter Bay.

POACEAE (GRAMINEAE) Grass Family

Contributed by Robert D. Dorn

Annual or perennial herbs. Lvs alt, sheathing, parallel-veined, usually with a membranous or hairy ligule at junction of sheath and blade on inner side. Fls bisex/uni, in spikelets, each spikelet usually consisting of 2 empty lower bracts (glumes) subtending 1 or more florets, each floret composed of 2 bracts (a lemma which usually has a midnerve, and a palea which usually lacks a midnerve) which subtend the flower. Ov superior. Per greatly reduced, S 3(1-6), P 2(1-3)*. Caryopsis (grain).

Several annual species including wheat, oats, and barley are commonly cultivated or are introduced in other seed, but these rarely, if ever, spread and persist more than a year. Oats (*Avena sativa* L.) and barley (*Hordeum vulgare* L.) have been found growing at the Kelly dump.

1 Spikelets of crowded scales subtending bulblets rather than flowers or seeds, the bulblets usually purplish; stems usually bulbous at base, densely tufted; leaf tips boat-shaped .... **Poa bulbosa**

1 Spikelets containing flowers or seeds; stems and leaves various

  2 Spikelets enclosed by a bur-like involucre bearing coalescent bristles forming spines; sandbur ...... **Cenchrus**

  2 Spikelets not as above, if bristles present, these not coalescent to form spines

    3 Spikelets sessile, forming terminal or lateral spikes, occasionally with the lower spikelets short-pediceled but then the glumes usually bristle-like or nearly so

      4 Spikelets with 1 perfect terminal floret and 2 opposite, sterile lemmas below .... **Phalaris**

4 Spikelets with all perfect florets, or with only 1 sterile lemma below, or with sterile lemmas all above the fertile
5 Spikelets dorsally flattened and falling entire, with 1 perfect terminal floret and 1 sterile lemma (which resembles a glume) or staminate floret below, without awns .... **Setaria**
5 Spikelets usually flattened from the sides, the florets often falling individually with the glumes persistent; spikelets usually with 1 or more perfect florets, with or without sterile or staminate florets, or sometimes the flowers all unisexual; lemmas or glumes awned or not
6 Spikes 1 or more, usually lateral or not directly continuous with the main axis; spikelets often on only 1 side of rachis
7 Plants with creeping rhizomes; glumes unequal in length .... **Spartina**
7 Plants lacking rhizomes; glumes equal in length .... **Beckmannia**
6 Spikes single and terminal; spikelets on opposite sides of rachis
8 Spikelets 1 flowered, the lemma much shorter than the glumes and awnless **Phleum**
8 Spikelets 2 or more flowered, or if 1 flowered, the lemma conspicuously awned
9 Spikelets mostly 1 per node ... **Agropyron**
9 Spikelets mostly 2 or more per node, at least at middle of spike, the lateral ones sometimes reduced to awns
10 Spikelets 3 per node, mostly 1 flowered, the lateral ones pediceled and usually reduced to awns . **Hordeum**
10 Spikelets 2 or more per node, 2 or more flowered, the lateral ones like the central one and usually sessile
11 Rachis usually continuous; glumes broad, or subulate and bristle-like; awns usually less

than 3 cm long .... **Elymus**

11 Rachis disarticulating when mature; glumes subulate, often bristle-like; awns usually over 3 cm long .... **Sitanion**

3 Spikelets, or most of them, with very short or long pedicels, the inflorescence a raceme or panicle which is sometimes spike-like; rarely with a single spikelet

12 Plants stout reeds to 4 m high with plume-like panicles; at least some leaves 9 mm or more wide; rachilla with long silky hairs as long as the lemmas, the hairs often inconspicuous in young flowers; spikelets several flowered, the florets sometimes poorly differentiated in very young spikelets; glumes shorter than lowest lemma; lowest lemmas mostly 9 mm or more long; usually in moist areas or in water .... **Phragmites**

12 Plants not as above

13 Spikelets usually dorsally flattened and falling entire, with 1 perfect terminal floret and usually 1 sterile lemma (which resembles a glume) or staminate floret below (1st glume sometimes minute)

14 Inflorescence appearing like a simple spike; spikelets subtended by long bristles .... **Setaria**

14 Inflorescence not appearing like a simple spike; spikelets not subtended by bristles .... **Panicum**

13 Spikelets usually flattened from the sides, the florets usually falling individually with the glumes persistent; spikelets with 1 or more perfect florets or the plants rarely with only staminate or pistillate flowers; sterile or staminate florets, if any, above the perfect or with 2 below the perfect

15 Spikelets with 1 perfect terminal floret and 2 sterile lemmas or staminate florets below, the sterile often reduced to linear lemmas with long hairs

16 Lower florets staminate, well developed; spikelets brown and shiny; inflorescence

an open panicle .... **Hierochloë**

16 Lower florets reduced to small scale-like or linear lemmas; spikelets green or yellow and dull; inflorescence a spike-like or contracted panicle .... **Phalaris**

15 Spikelets not as above, the sterile florets, if present, above the fertile

17 Spikelets mostly with 1 floret

18 Disjointing below the glumes, the entire spikelets falling (most evident on mature plants but joints near tip of pedicels often apparent in younger plants)

19 Glumes awned .... **Phleum**

19 Glumes not awned

20 Panicle spike-like, cylindrical; keel of glumes long-ciliate .... **Alopecurus**

20 Panicle open; keel or midnerve of glumes glabrous or scabrous

21 Plants annual; spikelets about as long as wide .... **Beckmannia**

21 Plants perennial; spikelets over twice as long as wide .. **Cinna**

18 Disjointing above the glumes, the glumes not falling with the florets

22 Lemma hardened, much more so than glumes at maturity, closely enveloping grain, often without evident nerves, terminally awned or nearly so, the awn sometimes deciduous and sometimes over 2 cm long

23 Awn persistent, strongly twisted and bent, often over 1 cm long; callus often sharp pointed, usually acuminate; glumes 6-25 mm long .... **Stipa**

23 Awn often deciduous, not strongly twisted, sometimes

bent, mostly 1 cm or less long; callus usually obtuse; glumes 4-8 mm long .... **Oryzopsis**

22 Lemma not hardened, loose around the grain, usually with 1 or more evident nerves, awned or not, the awns, when present, usually less than 2 cm long

24 Glumes (excluding awns) longer than the lemmas

25 Glumes strongly flattened and keeled, stiff-ciliate on keel, short-awned; panicle spike-like; lemma awnless .... **Phleum**

25 Glumes not as above; panicle open to spike-like; lemmas awned or not

26 Floret with a tuft of hairs at base from the callus, the hairs usually ¼ as long to as long as lemma; palea well developed

27 Lemma awned from back .... **Calamagrostis**

27 Lemma awned from tip .... **Muhlenbergia**

26 Floret lacking hairs at base or with very short hairs; palea often small or lacking .... **Agrostis**

24 Glumes (excluding awns) mostly shorter than the lemmas

28 Lemmas awned from tip (tip rarely bifid) or mucronate (at least some with midnerve slightly prolonged beyond body); glumes often awned .... **Muhlenbergia**

28 Lemmas awnless or awned from the back; glumes not awned

29 Plants annual . . **Muhlenbergia**

29 Plants perennial . . . . **Catabrosa**

17 Spikelets with 2 or more florets

30 Glumes, or at least 1 glume, as long as or longer than the lowest floret, usually as long as the spikelet; lemmas awned from the back or from a bifid tip or awnless

31 Spikelets mostly 9 mm or more long; rachilla usually not prolonged beyond terminal floret . . **Danthonia**

31 Spikelets 8 mm long or less; rachilla often prolonged beyond terminal floret

32 Glume dissimilar, the 2nd much wider than the 1st; spikelets 2-4 mm long . . . . **Sphenopholis**

32 Glumes usually relatively similar; spikelets 4-8 mm long

33 Lemmas awned from near or below middle . . . . . . **Deschampsia**

33 Lemmas awnless or awned from above middle

34 Lemmas with an exserted, usually geniculate awn . . . . **Trisetum**

34 Lemmas awnless or with a very short, straight awn

35 Ligules mostly 2.5-4 mm long; rachilla strongly bearded . **Trisetum**

35 Ligules mostly 0.5-2.5 mm long; rachilla glabrous to very short hairy

36 Pedicels mostly less than 3 mm long; upper floret rarely exceeding longest glume by more than 1 mm . . . . **Koeleria**

36 Pedicels, or some of them, usually over 3 mm long; upper floret usually exceeding longest glume by 1.5 mm or more . . . . **Poa**

30 Glumes mostly shorter than the lowest floret; lemmas awned from the tip or from a bifid tip or awnless

37 Plants with either all staminate or all pistillate flowers, sometimes with rudiments of the opposite sex in a normally developed flower

38 Plants with longer creeping rhizomes; sheaths usually long-hairy near throat; ligules usually with a fringe of hairs at tip; spikelets mostly 7-15 flowered . . . . **Distichlis**

38 Plants with rhizomes or not; sheaths not long-hairy; ligules mostly membranous; spikelets mostly 3-7 flowered

39 Panicle narrow and congested; some leaf blades often over 3.5 mm wide, often glaucous; plants short-rhizomatous . . . . **Leucopoa**

39 Panicle often open; leaf blades rarely over 3.5 mm wide, not glaucous; plants rhizomatous or not . . . . **Poa**

37 Plants with bisexual flowers

40 Plants annuals or tufted perennials; 1st glume much narrower than 2nd; spikelets mostly 2 flowered, 2-4 mm long, awnless, falling entire (the glumes not persistent) . . . . **Sphenopholis**

40 Plants without the above com-

bination of characters

41 Lemmas with 3 prominent nerves . . . . **Catabrosa**

41 Lemmas with 5 or more nerves or appearing nerveless

42 Spikelets crowded in 1 sided clusters at the ends of stiff, naked panicle branches; glumes usually hispid-ciliate on keel and sometimes on margins and nerves, otherwise mostly glabrous . . . . . **Dactylis**

42 Spikelets and glumes not as above

43 Stems usually bulbous at base in the soil; spikelets often tawny or purplish tinged; upper floret often sterile and lacking a palea; sheaths usually closed most of their length . . . . **Melica**

43 Stems usually not bulbous at base; spikelets often completely green; upper florets perfect, or if not, with paleas, or the floret reduced to a rudiment; sheaths often split most of their length

44 Lemmas mostly obtuse and scarious at tip, not awned, 5-9 nerved; glumes mostly 3 mm or less long; leaves not boat-shaped at tip or slightly so in drying

45 Second glume 1 nerved; styles well developed; nerves of lemmas prominent; plants of fresh water shores or moist woods **Glyceria**

45 Second glume usually 3 nerved; styles lacking or nearly so; nerves of lemmas prominent or obscure; plants of alkaline soil or fresh water areas .... **Puccinellia**

44 Lemmas sometimes acute at tip, scarious or not, often awned, if not awned, rarely over 5 nerved; glumes often over 3 mm long; leaves sometimes boat-shaped at tip

46 Lemmas awned or awn-tipped from a minutely or strongly bifid tip; spikelets usually 15 mm or more long .... **Bromus**

46 Lemmas often entire, pointed or obtuse, awnless or awned usually from the tip; spikelets mostly

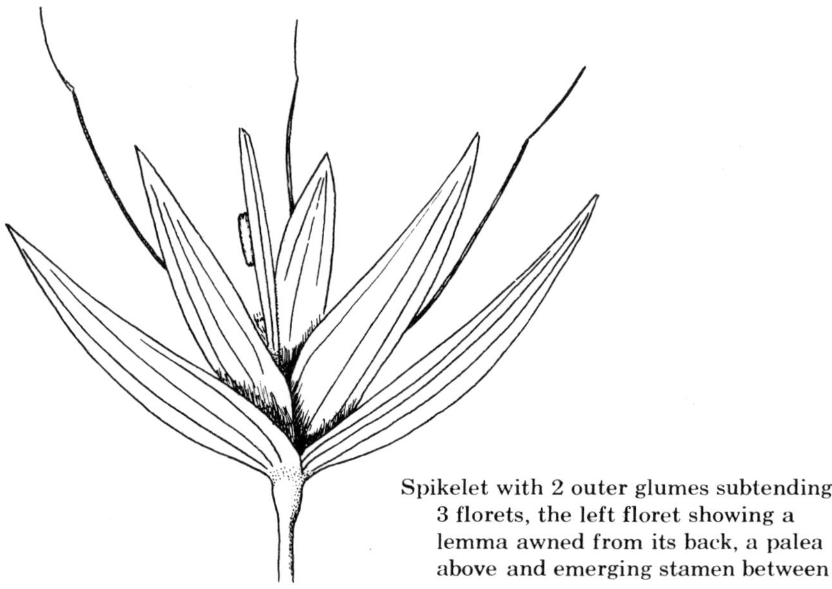

Spikelet with 2 outer glumes subtending 3 florets, the left floret showing a lemma awned from its back, a palea above and emerging stamen between

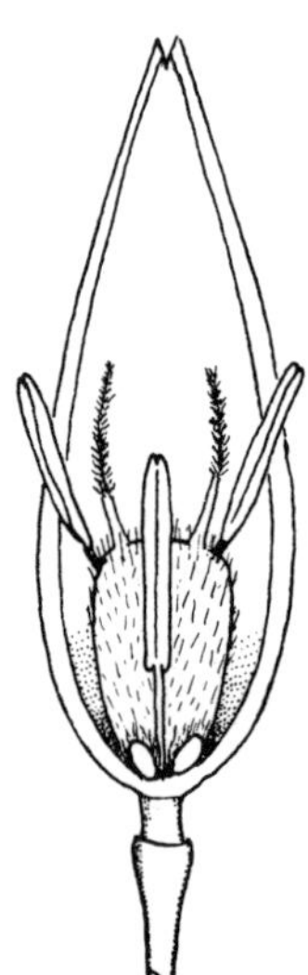

Ground plan of flower (X-sec.)

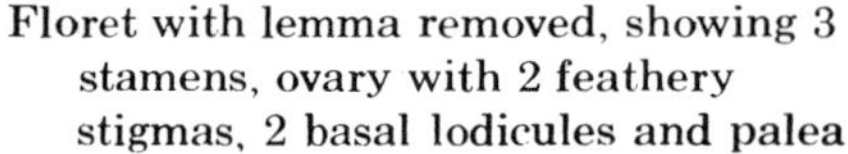

Floret with lemma removed, showing 3 stamens, ovary with 2 feathery stigmas, 2 basal lodicules and palea

Poaceae

less than 15 mm long

47 Lemmas awned, mostly rounded on back and with slender pointed tips .... **Festuca**

47 Lemmas awnless (midnerve rarely slightly extended), often keeled and blunt and scarious at tip .... **Poa**

AGROPYRON Wheatgrass

Spikelets several flowered, solitary or rarely in pairs, sessile; glumes mostly subequal, usually shorter than 1st lemma; lemmas 5-7 nerved, mostly acute or awned.

1 Plants with creeping rhizomes
  2 Leaf blades mostly flat, some 5-10 mm wide; awn, if present, straight .... **A. repens**
  2 Leaf blades either involute or much less than 5 mm wide; awn, if present, often divergent
    3 Glumes rigid, usually widest near base, often as long as 1st lemma, mostly 3-5 nerved, short-awned .... **A. smithii**
    3 Glumes not rigid, widest at or above middle, shorter than 1st lemma, mostly 5-7 nerved, acute to awn-tipped .... **A. dasystachyum**
1 Plants without creeping rhizomes
  4 Spikelets strongly divergent, much compressed and crowded, some at least 4 times as long as internodes of rachis .... **A. cristatum**
  4 Spikelets usually erect or ascending, not much compressed or crowded, mostly 3 times as long as internodes or less

5 Anthers 4-6 mm long; spikelets shorter to slightly longer than internodes of rachis; glumes acute or awn-tipped; lemmas often with a divergent awn .. .. **A. spicatum**

5 Anthers 1-3 mm long; spikelets mostly 2-3 times as long as internodes of rachis; glumes and lemmas various

6 Awn of lemma divergent; glumes narrowly lanceolate, long-awned; rachis tending to disarticulate when mature .... **A. scribneri**

6 Awn of lemma either lacking or mostly straight; glumes oblong-elliptic, awned or not; rachis not disarticulating .... **A. caninum**

1. A. caninum (L.) Beauv. ssp. **majus** (Vasey) Hitchc.

Spikelets mostly 1-2.5 times as long as internodes, the spikes mostly 8-20 cm long; auricles usually present; plants mostly of the lowlands extending into the mountains

Lemmas lacking awns or these to 5 mm long .... var. **majus**

Lemmas with awns (6) 10-30 mm long .... var. **unilaterale** (Vasey) Hitchc.

Spikelets mostly 2.5-3 times as long as internodes, the spikes mostly 4-8 cm long; auricles usually lacking; plants mostly of the mountains

Lemmas lacking awns or these to 6 mm long .... var. **latiglume** (Schribn. & Sm.) Hitchc.

Lemmas with awns mostly 8-20 mm long .... var. **andinum** (Scribn. & Sm.) Hitchc.

Var. **andinum** [*A. subsecundum* (Link) Hitchc.]. Frequent along streams and in moist woods.

Var. **latiglume** [*A. trachycaulum* (Link) Malte]. Frequent along Snake River.

Var. **majus.** Rare in the valleys.

Var. **unilaterale.** Rare in the valleys.

†**2. A. cristatum** (L.) Gaertn. Frequent along roads and in other disturbed areas.

3. A. dasystachyum (Hook.) Scribn. Infrequent with sagebrush or in open areas. Including *A. albicans* Scribn. & Sm.

4. A. repens (L.) Beauv. Infrequent in disturbed areas.

5. **A. scribneri** Vasey. Frequent, alpine and subalpine.
6. **A. smithii** Rydb. Infrequent in the valleys.
7. **A. spicatum** (Pursh) Scribn. & Sm. Frequent in the valleys.

AGROSTIS Bentgrass; Redtop

Spikelets 1 flowered; glumes usually equal or nearly so; lemma usually shorter than glumes, mostly 3 nerved, awned or not.

1 Palea evident, 2 nerved, over half as long as lemma
  2 Rachilla prolonged behind palea as a minute stub or bristle .... **A. thurberiana**
  2 Rachilla not prolonged
    3 Plants tufted, alpine, less than 18 cm high ... **A. humilis**
    3 Plants with rhizomes, rarely alpine, usually over 25 cm high .... **A. alba**
1 Palea lacking or a minute nerveless scale
  4 Panicle contracted, at least some lower branches spikelet-bearing near base
    5 Culms not over 20 cm high, in dense tufts with many basal leaves; leaf blades less than 2 mm wide .... **A. variabilis**
    5 Culms over 20 cm high, not in tufts with many basal leaves; leaf blades mostly over 2 mm wide .. .. **A. exarata**
  4 Panicle open at maturity, the lower branches not spikelet-bearing near base
    6 Panicle diffuse, the capillary branches branching toward the end .... **A. scabra**
    6 Panicle not diffuse, the branches branching at or below the middle
      7 Spikelets mostly 1.5-2 mm long; plants less than 30 cm high .... **A. idahoensis**
      7 Spikelets 2-3 mm long; plants usually over 30 cm high .... **A. oregonensis**

1. **A. alba** L.

Panicle open, mostly over 15 mm wide when pressed; plants predominantly rhizomatous .... var. **alba**

Panicle congested, mostly 15 mm or less wide when pressed; plants predominantly stoloniferous .... var. **palustris** (Huds.) Pers.

Var. **alba.** Frequent in aspen groves and in meadows.
Var. **palustris.** Infrequent in moist places.

2. A. **exarata** Trin. ssp. **minor** (Hook.) Hitchc. Frequent in open places.
3. A. **humilis** Vasey. Infrequent, alpine or subalpine.
4. A. **idahoensis** Nash. Infrequent in mountain meadows.
5. A. **oregonensis** Vasey. Rare in wet places.
6. A **scabra** Willd.

Panicle branches relatively long and ascending . var. **scabra**
Panicle branches relatively short and widely spreading
.... var. **geminata** (Trin.) Swallen

Var. **geminata.** Frequent in open places.
Var. **scabra.** Infrequent in open places.

7. A. **thurberiana** Hitchc. Rare in the mountains.
8. A. **variabilis** Rydb. Rare in the high mountains.

ALOPECURUS Foxtail

Panicle spike-like; spikelets 1 flowered; glumes equal, usually united at base, ciliate on keel; lemma about half as long to as long as glumes, 5 nerved, awned from below middle; palea lacking.

1 Spikelets densely woolly; panicle about 1 cm wide .. **A. alpinus**
1 Spikelets often hairy but not woolly; panicle less than 7 mm wide .... **A. aequalis**

1. **A. aequalis** Sobol. Frequent in water and wet areas.
2. **A. alpinus** Smith. Rare in moist meadows and along streams.

BECKMANNIA Sloughgrass

Spikelets 1 flowered, in 2 rows along 1 side of rachis; lemma about as long as glumes, 5 nerved, awnless but acuminate.

1. **B. syzigachne** (Steud.) Fern. Infrequent in water or wet places.

BROMUS Bromegrass; Chess

Spikelets several to many flowered, in a panicle; glumes unequal to subequal; lemmas 5-9 nerved, 2 toothed, awned from between the teeth or awnless.

1 Spikelets strongly flattened, the lemmas compressed-keeled; plants tufted perennials .... **B. carinatus**
1 Spikelets terete or somewhat flattened but the lemmas not compressed-keeled; plants annual or perennial

2 Plants perennial
  3 Creeping rhizomes present .... **B. inermis**
  3 Creeping rhizomes lacking
    4 Lemmas pubescent along the margin and on lower part of back, the upper part glabrous or nearly so
      5 Ligule 3-5 mm long; awn usually over 5 mm long .... **B. vulgaris**
      5 Ligule about 1 mm long; awn 3-5 mm long .... **B. ciliatus**
    4 Lemmas pubescent evenly over the back, usually more densely so along the lower part of margin .... **B. anomalus**
2 Plants annual
  6 Teeth of lemmas 2-3 mm long; awns mostly 10-15 mm long; 1st glume 1 nerved, 2nd glume usually 3 nerved .... **B. tectorum**
  6 Teeth of lemmas mostly less than 1 mm long; awns often less than 10 mm long; 1st glume 3-5 nerved, 2nd glume 5-9 nerved
    7 Awn straight; panicle branches stiffly spreading or ascending, usually not flexuous ... **B. commutatus**
    7 Awn flexuous, usually divergent when dry; panicle branches lax or flexuous .... **B. japonicus**

1. **B. anomalus** Rupr. ex Fourn. Frequent, woods and slopes.
2. **B. carinatus** Hook. & Arn.
   Leaf blades and sheaths glabrous or pubescent, the culms glabrous or hairy only near nodes; leaves mostly over 5 mm wide .... var. **carinatus**
   Leaf blades and sheaths densely pilose, the culms with shorter hairs; leaves mostly less than 5 mm wide .... var. **linearis** Shear

   Var. **carinatus** [*B. marginatus* Nees]. Frequent in mountain meadows.

   Var. **linearis.** Frequent in sagebrush.
3. **B. ciliatus** L. Frequent in moist places.

†4. **B. commutatus** Schrad. Frequent along roads.

5. **B. inermis** Leyss.
   First glume tapered from base; plants usually glabrous; ligules usually 1 mm or less long .. ssp. **inermis**

First glume widened above base; plants usually pubescent at least on lemmas and nodes; ligules mostly 1-2 mm long .... ssp. **pumpellianus** (Scribn.) Wagnon

Glumes glabrous; lemmas hairy mostly on lower half and along veins .... var. **pumpellianus**

Glumes pubescent; lemmas usually hairy throughout .... var. **tweedyi** (Scribn.) Hitchc.

Ssp. **inermis.** Frequent in aspen groves.

Ssp. **pumpellianus**

Var. **pumpellianus.** Infrequent in open places.

Var. **tweedyi.** Infrequent in woods and meadows.

†**6. B. japonicus** Thunb. Infrequent along roads.

†**7. B. tectorum** L. Cheatgrass. Frequent in disturbed areas.

8. B. vulgaris (Hook.) Shear. Rare in woods.

CALAMAGROSTIS Reedgrass

Spikelets 1 flowered, in open to spike-like panicles; glumes about equal; lemma shorter than glumes, awned from back, the callus bearing a tuft of hairs mostly ¼ as long as lemma to as long as lemma.

1 Awn exserted 1-4 mm beyond glumes, geniculate; glumes mostly 6-8 mm long .... **C. purpurascens**

1 Awn either included or scarcely longer than the glumes, straight or geniculate; glumes mostly shorter

2 Callus hairs rarely over half as long as lemma (hairs of rachilla sometimes longer); awn geniculate or straight

3 Sheaths, or some of them, pubescent on the collar, the hairs much longer than foliage hairs if present .... **C. rubescens**

3 Sheaths glabrous on the collar or the hairs the same as the foliage hairs

4 Awn attached toward base of lemma . **C. koelerioides**

4 Awn attached at or above middle of lemma .. .. **C. scopulorum**

2 Callus hairs mostly ⅔ as long as lemma to as long as lemma; awn straight

5 Panicle loose and usually open, mostly over 2 cm wide; awn delicate; leaf blades often over 4 mm wide, usually flat .... **C. canadensis**

5 Panicle contracted or spike-like, rarely over 3 cm wide; awn somewhat stout; leaf blades 1-4 mm wide, sometimes involute
6 Ligules of upper leaves 4-8 mm long .... **C. inexpansa**
6 Ligules of upper leaves 1-3.5 mm long ...... **C. stricta**

1. **C. canadensis** (Michx.) Beauv. (including *C. scribneri* Beal)
Collar hairy on at least some leaves .... var. **robusta** Vasey
Collar glabrous or scabridulous
Glumes mostly 3.8 mm or more long, acuminate .. .. var. **acuminata** Vasey
Glumes averaging less than 3.8 mm long, often acute .... var. **canadensis**

Var. **acuminata.** Rare in moist places.
Var. **canadensis.** Frequent in moist places.
Var. **robusta.** Rare in moist places.

2. **C. inexpansa** Gray. Frequent in moist places.
3. **C. koelerioides** Vasey. Rare, mountain meadows and slopes.
4. **C. purpurascens** R. Br. Rare, rocky places in the mountains.
5. **C. rubescens** Buckl. Frequent in woods or sometimes in open places.
6. **C. scopulorum** Jones. Rare in moist places.
7. **C. stricta** (Timm) Koeler [*C. neglecta* (Ehrh.) Gaertn. et al.]. Rare in wet places.

CATABROSA Brookgrass

Spikelets mostly 2 flowered, in a panicle; glumes unequal, shorter than lower floret, truncate or erose at tip.

1. **C. aquatica** (L.) Beauv. Infrequent in shallow water or wet places.

CENCHRUS Sandbur

Spikelets solitary or few in a raceme, enclosed by a spiny bur.

1. **C. longispinus** (Hack.) Fern. Rare in Snake River Canyon.

CINNA Woodreed

Spikelets 1 flowered, in a panicle; glumes subequal; lemma about as long as glumes, awned or not.

1. **C. latifolia** (Trevir. ex Goepp.) Griseb. Rare in moist woods and along streams.

DACTYLIS Orchard Grass

Spikelets several flowered, nearly sessile in dense, 1 sided fascicles; glumes unequal; lemmas mucronate or short-awned, ciliate on keel.

†1. **D. glomerata** L. Frequent in meadows and disturbed areas.

DANTHONIA Oatgrass

Spikelets several flowered, in open or spike-like panicles; glumes subequal, mostly exceeding uppermost floret; lemmas with a flat, twisted, geniculate awn arising from between 2 terminal teeth.

1 Lemmas pilose on back, sometimes sparsely so ...... **D. spicata**
1 Lemmas glabrous on back, pilose on margin only
  2 Panicle narrow, the pedicels mostly appressed to rachis; spikelets mostly 4-10 per panicle .... **D. intermedia**
  2 Panicle open, the pedicels mostly spreading or reflexed; spikelets mostly 1-5 per panicle
    3 Panicle usually with a single spikelet, rarely 2 or 3; plants less than 3 dm high .... **D. unispicata**
    3 Panicle with few to several spikelets; plants usually over 3 dm high .... **D. californica**

1. **D. californica** Boland.
   Leaves pilose .... var. **americana** (Scribn.) Hitchc.
   Leaves glabrous or nearly so .... var. **californica**
   Var. **americana.** Rare in meadows and open woods.
   Var. **californica.** Rare, mountain meadows and slopes.
2. **D. intermedia** Vasey. Frequent in meadows and woods.
3. **D. spicata** (L.) Beauv. ex R. & S. var. **pinetorum** Piper. Rare in woods and open places.
4. **D. unispicata** (Thurb.) Munro ex Macoun. Frequent with sagebrush.

DESCHAMPSIA Hairgrass

Spikelets 2 flowered, in a panicle; glumes subequal, exceeding the florets or nearly so; lemmas awned from near or below middle.

1 Plants annual; leaves few .... **D. danthonioides**
1 Plants perennial; leaves usually numerous
  2 Panicle narrow, the branches mostly appressed to rachis or nearly so; leaf blades 0.5-1.5 mm wide . **D. elongata**

2 Panicle open, the branches spreading or ascending; leaf blades often over 1.5 mm wide
3 Awns mostly 2.5-3 mm long, from near middle of lemma; anthers 0.8-1.2 mm long .... **D. atropurpurea**
3 Awns mostly 3-4 mm long, from near base of lemma; anthers 1.2-2.2 mm long .... **D. cespitosa**

1. **D. atropurpurea** (Wahl.) Scheele. Frequent in mountain meadows and canyons.
2. **D. cespitosa** (L.) Beauv. Frequent in moist places.
3. **D. danthonioides** (Trin.) Munro ex Benth. Rare in open places.
4. **D. elongata** (Hook.) Munro ex Benth. Rare in open areas in the mountains.

DISTICHLIS Saltgrass

Spikelets 7-15 flowered, in a panicle; flowers unisexual.

1. **D. stricta** (Torr.) Rydb. Rare, alkaline area east of Elk Ranch Reservoir.

ELYMUS Wild Rye

Spikelets mostly 2-6 flowered, usually in pairs, mostly sessile; glumes subequal, sometimes bristle-like; lemmas obscurely 5 nerved, awned or awn-tipped.

1 Glumes somewhat bristle-like, not broadened above base, the nerves obscure .... **E. cinereus**
1 Glumes lanceolate or narrower, broadened above base, strongly nerved,
2 Glumes relatively thin, not indurate at base .... **E. glaucus**
2 Glumes firm, usually strongly indurate at base .... **E. canadensis**

1. **E. canadensis** L. Rare in open places especially where moist.
2. **E. cinereus** Scribn. & Merr. Frequent in open places.
3. **E. glaucus** Buckl. Frequent in open places.

FESTUCA Fescue

Spikelets few to several flowered, rarely 1 flowered in some spikelets, in a panicle; glumes unequal, mostly shorter than or equalling 1st lemma; lemmas awned from tip, 5 nerved, the nerves often obscure.

1 Leaf blades flat, usually over 3 mm wide .... **F. subulata**
1 Leaf blades involute, or if flat, less than 3 mm wide
  2 Culms either decumbent at the usually red or purple, fibrillose base or from rhizomes .... **F. rubra**
  2 Culms erect, without rhizomes, often not red, purple, or fibrillose at base
    3 Culms usually over 30 cm high; panicles 1-2 dm long, mostly open; anthers 2-4 mm long .... **F. idahoensis**
    3 Culms mostly less than 30 cm high; panicles mostly less than 1 dm long, mostly narrow; anthers 0.3-2 mm long .... **F. ovina**

1. **F. idahoensis** Elmer. Frequent with sagebrush.
2. **F. ovina** L.

Plants mostly 5-20 cm high, alpine and subalpine .. .. var. **brevifolia** Wats.
Plants mostly over 20 cm high, mostly below subalpine .... var. **rydbergii St.-Yves**

Var. **brevifolia.** Frequent in high mountain meadows.
Var. **rydbergii.** Infrequent in open areas.

3. **F. rubra** L. Rare in meadows and swamps.
4. **F. subulata Trin.** Rare in moist thickets.

GLYCERIA Mannagrass

Spikelets few to many flowered, in panicles; glumes unequal,
1 nerved; lemmas scarious at tip, 5-9 nerved.

1 Spikelets linear, nearly terete, usually 1 cm long or more; panicle narrow and erect .... **G. borealis**
1 Spikelets ovate or oblong, somewhat compressed, usually 6 mm or less long; panicle usually nodding
  2 Leaf blades mostly 2-5 mm wide; 1st glume 0.5-1 mm long; upper ligules closed in front (sometimes opened by drying) .... **G. striata**
  2 Leaf blades mostly 6 mm or more wide; 1st glume about 1 mm or more long; upper ligules usually open in front
    3 Ligules pubescent-scabridulous; 1st glumes averaging about 1 mm long .... **G. elata**
    3 Ligules glabrous; 1st glumes averaging about 1.5 mm long .... **G. grandis**

1. **G. borealis** (Nash) Batchelder. Frequent in moist places.
2. **G. elata** (Nash ex Rydb.) Jones. Frequent in moist places.
3. **G. grandis** Wats. Rare in wet places.
4. **G. striata** (Lam.) Hitchc. var. **stricta** (Scribn.) Fern. Frequent in moist places.

HIEROCHLOË Sweetgrass

Spikelets with 1 terminal perfect floret and 2 staminate florets below, in a panicle; lemmas about as long as glumes, awnless or nearly so.

1. **H. odorata** (L.) Beauv. Infrequent in moist meadows.

HORDEUM Barley

Spikelets usually 1 flowered, mostly 3 at each node, the central one sessile, the lateral ones usually pediceled; lateral spikelets usually imperfect, sometimes reduced to bristles; glumes awn-like or awned; lemmas awned or awn-tipped.

1 Awns 1.8-8 cm long .... **H. jubatum**
1 Awns 1.5 cm long or less .... **H. brachyantherum**

1. **H. brachyantherum** Nevski. Frequent in open, mostly moist places.

†2. **H. jubatum** L. Frequent in disturbed areas.

KOELERIA Junegrass

Spikelets 2-4 flowered, in a usually spike-like panicle; glumes unequal in width; lowest lemma usually slightly longer than glumes, acute or short-awned.

1. **K. macrantha** (Ledeb.) Shultes [*K. cristata* (L.) Pers. of authors]. Frequent with sagebrush.

LEUCOPOA Western Grass

Spikelets 3-5 flowered, in a narrow panicle, the flowers unisexual; glumes unequal; lemmas acute or acuminate.

1. **L. kingii** (Wats.) Weber. Infrequent, slopes and open woods. [*Hesperochloa kingii* (Wats.) Rydb.]

MELICA Melicgrass; Oniongrass

Spikelets 2 to several flowered, in a panicle; glumes somewhat unequal, shorter than lowest floret, 3-7 nerved; lemmas awned or not; base of culm often swollen.

1 Lemmas awned or long-tapering to a pointed tip
  2 Lemmas awned from a bifid tip .... **M. smithii**
  2 Lemmas acute or acuminate .... **M. subulata**
1 Lemmas mostly obtuse, awnless
  3 Pedicels mostly flexuous or recurved; 1st glume 3.5-5.5 mm long .... **M. spectabilis**
  3 Pedicels mostly straight; 1st glume 6-9 mm long **M. bulbosa**

1. **M. bulbosa** Geyer ex Porter & Coult. Rare in meadows.
2. **M. smithii** (Porter ex Gray) Vasey. Rare in moist woods.
3. **M. spectabilis** Scribn. Frequent in meadows and woods.
4. **M. subulata** (Griseb.) Scribn. Rare in meadows and woods.

MUHLENBERGIA Muhly

Spikelets 1 flowered or occasionally 2 flowered, in a panicle; lemmas usually awned from tip, the awn sometimes minute or rarely lacking.

1 Plants annual, the culms rarely decumbent and rooting at the nodes and appearing perennial; glumes 1 mm long or less .... **M. filiformis**
1 Plants perennial; glumes often over 1 mm long
  2 Panicle diffuse, the spikelets very remote on long pedicels or panicle branches .... **M. asperifolia**
  2 Panicle narrow and condensed, the spikelets crowded on short pedicels
    3 Hairs at base of floret about as long as body of lemma .... **M. andina**
    3 Hairs at base of floret not more than half as long as lemma
      4 Leaf blades mostly involute but sometimes flat, 1.2 mm wide or less .... **M. richardsonis**
      4 Leaf blades flat, mostly 2-7 mm wide ....**M. racemosa**

1. **M. andina** (Nutt.) Hitchc. Frequent in moist places.
2. **M. asperifolia** (Mey. ex Nees & Mey.) Parodi. Rare in moist, usually alkaline areas.
3. **M. filiformis** (Thurb. ex Wats.) Rydb. Frequent in moist places.
4. **M. racemosa** (Michx.) B. S. P. Frequent in meadows and thickets.

5. **M. richardsonis** (Trin.) Rydb. Frequent in open places.

ORYZOPSIS Ricegrass

Spikelets 1 flowered, in a panicle; glumes subequal; lemma about as long as glumes or shorter, usually awned, the awn deciduous.

1 Pubescence on lemma long and silky .... **O. hymenoides**
1 Pubescence on lemma short and appressed
  2 Spikelets, excluding awn, 6-9 mm long; leaf blades flat .... **O. asperifolia**
  2 Spikelets, excluding awn, 5.5 mm long or less; leaf blades involute or subinvolute .... **O. exigua**

1. **O. asperifolia** Michx. Rare in woods.
2. **O. exigua** Thurb. Infrequent, woods and slopes.
3. **O. hymenoides** (R. & S.) Ricker ex Piper. Infrequent with sagebrush.

PANICUM

Spikelets compressed dorsiventrally, in panicles; glumes very unequal, the 1st often minute; sterile lemma simulating a 3rd glume; margins of fertile lemma inrolled over an enclosed palea.

1 Plants annual; 1st glume 1-1.5 mm long .... **P. capillare**
1 Plants perennial; 1st glume 0.2-0.5 mm long ...... **P. thermale**

1. **P. capillare** L. var. **occidentale** Rydb. Rare in disturbed areas.
2. **P. thermale** Boland. Rare around hot springs.

PHALARIS Canary Grass

Spikelets with 1 terminal perfect floret and 2 much smaller, scale-like or linear, sterile lemmas below, in a narrow or spikelike panicle; glumes equal; fertile lemma shorter than glumes.

1 Plants annual; panicle 2-4 cm long .... **P. canariensis**
1 Plants perennial; panicle 6 cm or more long .. **P. arundinacea**

1. **P. arundinacea** L.
   Leaf blades white-striped .... var. **picta** L.
   Leaf blades not white-striped ....var. **arundinacea**

Var. **arundinacea.** Infrequent in wet places.

Var. **picta.** Rare among willows ½ mile north of Jackson Lake Lodge.

†**2. P. canariensis** L. Rare in moist, disturbed places.

PHLEUM Timothy

Spikelets 1 flowered, in a spike-like panicle; glumes equal, awned, ciliate on keel; lemma shorter than glumes, awnless.

1 Panicle long-cylindric, usually over 5 times as long as wide; culms usually bulbous at base .... **P. pratense**
1 Panicle ovoid or oblong, usually not over 4 times as long as wide; culms not bulbous at base .... **P. alpinum**

1. P. alpinum L. Frequent in canyons and on mountain slopes.
2. P. pratense L. Frequent in meadows and disturbed areas.

PHRAGMITES Common Reed

Spikelets several flowered, in a panicle; florets exceeded by hairs of rachilla; glumes unequal; lemmas long-acuminate or awned.

1. P. australis (Cav.) Trin. ex Steud. [*P. communis* Trin.]. Rare in water, north end of String Lake.

POA Bluegrass

Spikelets 2 to several flowered, in a panicle; glumes somewhat unequal; lemmas awnless, mostly 5 nerved.

1 Plants annual but often densely clustered, lacking remains of old culms, mostly 25 cm or less high .... **P. annua**
1 Plants perennial, usually with remains of old culms, often over 25 cm high
  2 Creeping rhizomes present, the culms often densely tufted also
    3 Culms strongly flattened, 2 edged .... **P. compressa**
    3 Culms terete or slightly flattened, not 2 edged
      4 Plants often dioecious, mostly pistillate; lower sheaths minutely retrorsely pubescent and usually purplish .... **P. nervosa**
      4 Plants usually perfect-flowered; lower sheaths not retrorsely pubescent, usually green
        5 Lemmas with tangled cobwebby hairs at base

6 Lemmas glabrous or scabrous on internerves, 3-4 mm long; mostly middle and lower elevations .... **P. pratensis**
6 Lemmas pubescent on internerves at least on lower half, 4-5 mm long; mostly alpine and subalpine, rarely lower .... **P. grayana**
5 Lemmas lacking cobwebby hairs at base
7 Plants alpine or subalpine .... **P. grayana**
7 Plants below subalpine
8 Panicle contracted, the branches ascending or appressed .... **P. arida**
8 Panicle open, the branches spreading or reflexed
9 Lemmas puberulent to glabrous; anthers 2.5-3 mm long; panicle branches mostly nodding .... **P. curta**
9 Lemmas villous at least on lower nerves; anthers 1.4-2.5 mm long; panicle branches spreading to ascending .... **P. grayana**
2 Creeping rhizomes lacking (culms sometimes decumbent and rooting)
10 Florets usually converted into bulblets with a dark purple base; culms bulbous at base .... **P. bulbosa**
10 Florets normal; culms not bulbous at base
11 Spikelets little compressed, the lemmas convex on back, the keels lacking or mostly obscure
12 Lemmas crisp-puberulent on back especially on nerves (sometimes obscure or only at very base)
13 Panicle open, the lower branches naked at base, mostly spreading .... **P. gracillima**
13 Panicle contracted, the branches appressed or somewhat spreading at anthesis, the lower branches often with spikelets near base
14 Culms slender, averaging under 30 cm high, with numerous short innovations at base; basal leaf blades mostly less than 1.5 mm wide and 5 cm long; spikelets often purplish tinged

15 Plants usually of dry foothills, flowering in spring, often strongly purplish tinged all over .... **P. sandbergii**

15 Plants montane to alpine, often in open rocky areas, flowering mostly in summer, rarely purplish tinged all over .... **P. incurva**

14 Culms stout, averaging over 40 cm high; innovations usually not numerous; basal leaf blades mostly 1-3 mm wide and over 5 cm long; spikelets sometimes purple banded but usually not purplish tinged

16 Ligules of upper leaves mostly 3-7 mm long .... **P. scabrella**

16 Ligules of upper leaves 1-3 mm long .... **P. juncifolia**

12 Lemmas glabrous or minutely scabrous, not crisp-puberulent

17 Sheaths scaberulous; ligules on upper leaves about 4 mm long, decurrent ...... **P. nevadensis**

17 Sheaths glabrous; ligules shorter, usually not decurrent .... **P. juncifolia**

11 Spikelets distinctly compressed, the glumes and lemmas usually strongly keeled

18 Lemmas with tangled cobwebby hairs as base (sometimes scant or obscure in *P. interior*)

19 Lower panicle branches distinctly reflexed at maturity, mostly capillary with the spikelets borne near the ends; anthers less than 1 mm long .... **P. reflexa**

19 Lower panicle branches not reflexed, capillary or not, sometimes with spikelets borne near base; anther length variable

20 Lower panicle branches in pairs, elongate, capillary, with a few spikelets near the ends; anthers less than 1 mm long

21 Sheaths retrorsely scabrous to glabrous; spikelets usually green; plants not alpine .... **P. tracyi**

21 Sheaths glabrous or scabridulous; spikelets often purplish; plants often

alpine or subalpine .... **P. leptocoma**

20 Lower panicle branches often more than 2, or if not, not capillary and elongate; anthers often over 1 mm long

22 Culms 2-5 dm high, densely tufted; ligules 0.5-1.5 mm long; panicle 5-15 cm long

23 Second glume usually 2-3 mm long; mostly below subalpine **P. interior**

23 Second glume usually 3-4.5 mm long; mostly alpine and subalpine .... **P. pattersonii**

22 Culms 3-12 dm high, loosely tufted; ligules 1.5-5 mm long; panicle 12-30 cm long

24 Ligules 1.5-3 mm long; lemmas 3-5 mm long .... **P. tracyi**

24 Ligules 3-5 mm long; lemmas 1.5-3 mm long .... **P. palustris**

18 Lemmas not cobwebby at base

25 Lemmas glabrous or sometimes uniformly puberulent .... **P. cusickii**

25 Lemmas conspicuously pubescent on keel or marginal nerves or both, sometimes also on internerves but these hairs shorter

26 Leaf blades folded or involute, firm, rather stiff; usually lower than subalpine .... **P. fendleriana**

26 Leaf blades flat, or if involute, rather lax or soft; usually alpine or subalpine

27 Panicle usually about as broad as long, pyramidal; leaf blades mostly over 1.5 mm wide .... **P. alpina**

27 Panicle longer than broad, oblong; leaf blades mostly 1.5 mm or less wide

28 Second glume 3-4.5 mm long, extending to about tip of 1st lemma .... **P. pattersonii**

28 Second glume 2.5-3.5 mm long, exceeded by 1st lemma .... **P. rupicola**

1. **P. alpina** L. Infrequent, mountain meadows or talus.
†2. **P. annua** L. Frequent in meadows and open woods.
3. **P. arida** Vasey. Rare in meadows.
†4. **P. bulbosa** L. Frequent in disturbed areas and meadows.
5. **P. compressa** L. Rare in disturbed areas and meadows.
6. **P. curta** Rydb. Rare in moist, shaded areas.
7. **P. cusickii** Vasey.

Basal leaf blades filiform and involute, mostly 1 mm or less wide, often over 10 cm long; upper blades usually rolled or folded and mostly 1 mm wide or less .... var. **cusickii**

Basal leaf blades often flat, usually averaging 1 mm or more wide and less than 10 cm long; upper blades flat or folded, mostly 2-3.5 mm wide .... var. **epilis** (Scribn.) Hitchc.

Var. **cusickii.** Frequent with sagebrush.
Var. **epilis.** Infrequent in meadows.

8. **P. fendleriana** (Steud.) Vasey. Frequent, meadows and slopes.
9. **P. gracillima** Vasey. Rare in rocky places.
10. **P. grayana** Vasey [*P. arctica* R. Br. of authors]. Infrequent in mountain meadows mostly above timberline.
11. **P. incurva** Scribn. & Wms. Infrequent, open woods to alpine.
12. **P. interior** Rydb. Frequent, woods, slopes, and meadows.
13. **P. juncifolia** Scribn. [*P. ampla* Merr.]. Frequent in woods, meadows and on slopes.
14. **P. leptocoma** Trin. Frequent in bogs and thickets.
15. **P. nervosa** (Hook.) Vasey var. **wheeleri** (Vasey) Hitchc. Frequent, open woods and slopes.
16. **P. nevadensis** Vasey ex Scribn. Rare in moist places.
17. **P. palustris** L. Frequent in moist places.
18. **P. pattersonii** Vasey. Infrequent, alpine or subalpine.
19. **P. pratensis** L. Frequent in open places.
20. **P. reflexa** Vasey & Scribn. Infrequent in moist places in the mountains.
21. **P. rupicola** Nash ex Rydb. Rare on rocky slopes in the high mountains.
22. **P. sandbergii** Vasey. Infrequent with sagebrush.
23. **P. scabrella** (Thurb.) Benth. ex Vasey [*P. canbyi* (Scribn.) Piper]. Frequent in open woods, meadows, and on slopes.
24. **P. tracyi** Vasey. Rare in moist places.

PUCCINELLIA Alkalai Grass; False Manna

Spikelets several flowered, in a panicle; glumes unequal, shorter than 1st lemma, 1 or 3 nerved; lemmas 5-7 nerved, usually scarious at tip.

1 Nerves of lemmas mostly obscure; ligules rarely over 3 mm long; plants mostly of alkaline areas ...... **P. nuttalliana**
1 Nerves of lemmas prominent, usually raised from surface; ligules 3 mm or more long; plants mostly of fresh water areas
 2 Leaf blades 1-3 mm wide; culms weak and decumbent .... **P. fernaldii**
 2 Leaf blades, or some of them, over 4 mm wide; culms stout and erect to ascending .... **P. pauciflora**

1. **P. fernaldii** (Hitchc.) Voss [*Torreyochloa fernaldii* (Hitchc.) Church]. Rare 1½ miles south of Moose usually in water.
2. **P. nuttalliana** (Schult.) Hitchc. Rare in moist, usually alkaline places east of Elk Ranch Reservoir.
3. **P. pauciflora** (Presl) Munz [*Torreyochloa pauciflora* (Presl) Church]. Frequent in moist places or in water. Our plants are referable to var. **holmii** (Beal) Hitchc.

SETARIA Bristlegrass

Annuals with the spikelets dorsally flattened and subtended by 1 to several bristles, awnless, and in a spike-like panicle; 1st glume usually about half the spikelet length or less; 2nd glume longer than 1st; sterile lemma present below the usually transversely rugose, fertile one.

†1. **S. viridis** (L.) Beauv. Rare in disturbed places.

SITANION Squirreltail

Spikelets mostly 2 to several flowered, sessile and usually 2 per node; glumes and lemmas awned, the rachis disarticulating when mature.

1. **S. hystrix** (Nutt.) Smith
 Glumes 4, simple; sterile glume-like lemmas lacking .... var. **brevifolium** (Smith) Hitchc.
 Glumes apparently 5 or more, simple to trifid; some sterile glume-like lemmas present .... var. **hystrix**

Var. **brevifolium.** Rare, disturbed areas, woods, and slopes.
Var. **hystrix.** Frequent, disturbed areas, woods, and slopes.

SPARTINA Cordgrass

Spikelets 1 flowered, sessile on 1 side of rachis; glumes keeled, unequal, the 1st about half as long as 2nd; lemma keeled, awnless.

1. S. gracilis Trin. Rare in moist alkaline areas.

SPHENOPHOLIS Wedgegrass

Spikelets mostly 2 flowered, in a panicle; glumes unequal, the 1st narrow and 1 nerved, the 2nd broad and 3-5 nerved; lemmas awnless.

1 Panicle dense, usually spike-like; 2nd glume about 1½ times as long as wide or less .... **S. obtusata**
1 Panicle loose, not spike-like; 2nd glume almost 3 times as long as wide .... **S. intermedia**

1. S. intermedia (Rydb.) Rydb. Rare in moist places.
2. S. obtusata (Michx.) Scribn. Infrequent in moist places.

STIPA Needlegrass

Spikelets 1 flowered, in a panicle; glumes acute to aristate; lemma usually strongly convolute, hardened, awned, the awn twisted below.

1 Glumes 15 mm or more long; lemmas 8-12 mm long; awns mostly 7-15 cm long .... **S. comata**
1 Glumes 13 mm or less long; lemmas 4-9 mm long; awns mostly less than 6 cm long
  2 Awns plumose toward base .... **S. occidentalis**
  2 Awns not plumose toward base
    3 Panicle open, the branches spreading or ascending and spikelet-bearing near tip .... **S. richardsonii**
    3 Panicle narrow, the branches appressed and often spikelet-bearing near the base
      4 Sheaths, at least the lowermost, pubescent .... **S. williamsii**
      4 Sheaths glabrous or villous only at the throat or on margins
        5 Sheaths villous at the throat; callus broad and blunt; lower nodes of panicle villous ... **S. viridula**

5 Sheaths not villous at throat or only slightly so; callus narrow and sharp-pointed; nodes of panicle glabrous or nearly so
6 Awns mostly over 2 cm long; lemmas mostly 6-7 mm long .... **S. columbiana**
6 Awns mostly less than 2 cm long; lemmas mostly 4-5 mm long .... **S. lettermanii**

1. **S. columbiana** Macoun. Frequent in meadows and woods.
2. **S. comata** Trin. & Rupr. Rare with sagebrush.
3. **S. lettermanii** Vasey. Frequent in sagebrush, woods, and on slopes.
4. **S. occidentalis** Thurb. ex Wats. Rare in sagebrush and open woods.
5. **S. richardsonii** Link. Frequent in woods and meadows.
6. **S. viridula** Trin. Rare with sagebrush.
7. **S. williamsii** Scribn. Rare in sagebrush and woods.

TRISETUM

Spikelets usually 2 or 3 flowered, sometimes more, in an open or spike-like panicle; glumes somewhat unequal, the 2nd usually exceeding the 1st floret; lemmas 2, toothed at tip, awned from the back or awnless.

1 Awns included in glumes or nearly so, 2 mm long or less, straight, or awns lacking .... **T. wolfii**
1 Awns exserted from glumes, 3 mm long or more, usually bent
2 Panicle dense, spike-like, sometimes interrupted below; plants densely tufted .... **T. spicatum**
2 Panicle loose and open to contracted but not spike-like; plants in small tufts or solitary .... **T. montanum**

1. **T. montanum** Vasey. Repeatedly reported from the area but all specimens I have seen so identified are *T. spicatum*.
2. **T. spicatum** (L.) Richt. Frequent in woods and on alpine slopes.
3. **T. wolfii** Vasey. Infrequent in sagebrush and meadows.

POTAMOGETONACEAE Pondweed Family

Aquatic herbs, slender rhizomes, stem jointed. Lvs alt/opp, 2 ranked, sheathing at the base. Infl axillary spikes. Fls bisex, ov superior. Per 4 valvate segments, S 4, P 4. Fruit drupaceous.

POTAMOGETON Pondweed

Characteristics of the family. The identification of species in this genus is difficult because immature plants will have neither flowers nor fruits. Hybridization among species is also a complicating factor.

1 Leaves below surface of water with stipules attached to the base of the leaf, forming a sheath around the stem
  2 Stigma attached to a short style which becomes a short beak on the fruit .... **P. pectinatus**
  2 Stigma attached without style, fruit not beaked **P. filiformis**
1 Leaves below the surface with stipules free of the rest of the leaf
    3 Stems flattened and more or less winged . **P. zosteriformis**
    3 Stems nearly round in cross-section
      4 Leaves clasping at the base, all submersed .... **P. richardsonii**
      4 Leaves not clasping at the base, often dimorphic
        5 Leaf bases with paired, spherical, yellowish glands
          6 Stipules white, strongly fibrous, shredding into persistent fibers .... **P. friesii**
          6 Stipules not white or fibrous, usually deciduous .... **P. pusillus**
        5 Leaf bases lacking glands
            7 Leaves all submerged, linear, less than 3 mm broad .... **P. foliosus**
            7 Leaves of two kinds, often more than 4 mm broad
              8 Leaves below surface very narrow, less than 2 mm broad and over 10 cm long .... **P. natans**
              8 Leaves below surface more than 2 mm broad, often less than 10 cm long
                9 Plants usually reddish; floating leaves (if any) not different from leaves below surface .... **P. alpinus**
                9 Plants greenish; floating leaves different from leaves below surface .... **P. gramineus**

Potamogeton gramineus Pondweed

Potamogetonaceae

1. **P. alpinus** Balbis var. **tenuifolius** (Raf.) Ogden. Frequent in ponds of the valley.
2. **P. filiformis** Pers. Common in ponds in the north end of the valley.
3. **P. foliosus** Raf. Infrequent in Swan Lake.
4. **P. friesii** Rupr. Frequent in ponds along the Snake River.
5. **P. gramineus** L. Frequent in ponds along the Snake River; most variable species.
6. **P. natans** L. Infrequent in Swan Lake.
7. **P. pectinatus** L. Not seen in G.T.N.P., but reported in Y.N.P.
8. **P. pusillus** L. Frequent in quiet water of Oxbow Bend of Snake River, the most common species in N. America.
9. **P. richardsonii** (Benn.) Rydb. Common in ponds of the valley.
10. **P. zosteriformis** Fern. Infrequent in Swan Lake.

SPARGANIACEAE Bur-reed Family

Aquatic herbs. Lvs alt, sheathing at the base. Fls unisex, globose heads, ov superior. Per 3-6 scales, S (usually) 5, P 1-2*. Fruit nut-like.

SPARGANIUM Bur-reed

Only one genus with the characteristics of family.

1 Leaves mostly over 5 mm broad; pistillate heads usually more than 2 cm thick at maturity .... **S. emersum**
1 Leaves mostly less than 5 mm broad; pistillate heads 1-2 cm thick at maturity .... **S. angustifolium**

1. **S. angustifolium** Michx. Infrequent in beaver ponds along the Snake River and ponds near Colter Bay.
2. **S. emersum** Rehmann. Infrequent in shallow ponds of the valley. [*S. simplex* Huds.].

TYPHACEAE Cat-tail Family

Rhizomatous herbs. Lvs alt, linear and mostly basal. Infl of 2 unisex superimposed spikes, ov superior. Per slender hairs, S 2-5, P 1. Fruit dry eventually dehiscent.

TYPHA Cat-tail

Characteristics of the family.

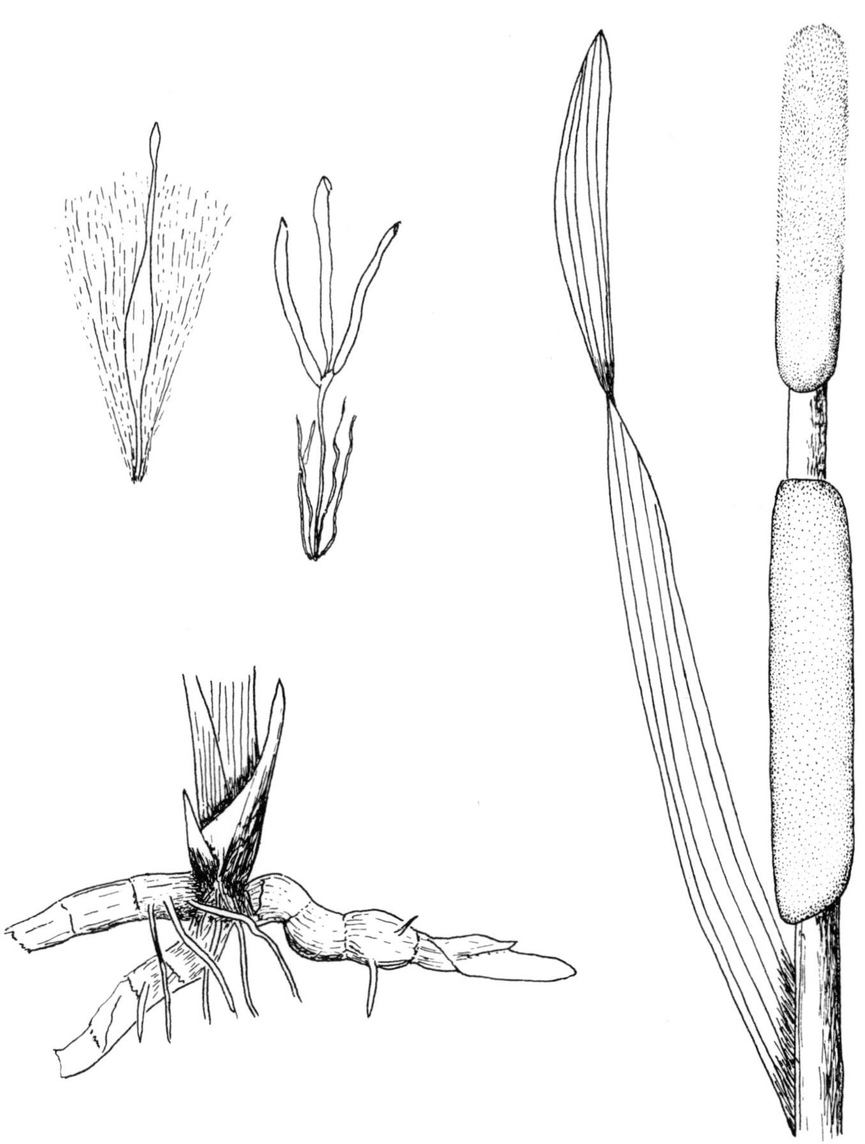

Typha angustifolia Cat-tail **Typhaceae**

1 Staminate and pistillate portions of the spike contiguous; leaves mostly 8-20 mm broad .... **T. latifolia**

1 Staminate and pistillate portions of the spike separated by a length of naked axis; leaves mostly 5-10 mm broad .... **T. angustifolia**

1. **T. angustifolia** L. Quiet to slow-moving water. Reported in Y.N.P.

1. **T. latifolia** L. Locally common in shallow ponds east of Jackson Lake.

ZANNICHELLIACEAE Horned Pondweed Family

Aquatic herbs, rhizomes. Lvs opp, entire, with free membranous stipules. Fls imperfect, naked, axillary, S 1, P 2-9. Fruit achene. Our species monoecious.

ZANNICHELLIA Horned Pondweed

Characteristics of the family.

1. **Z. palustris** L. Submersed in fresh to brackish water. Not seen in G.T.N.P., but reported in Y.N.P.

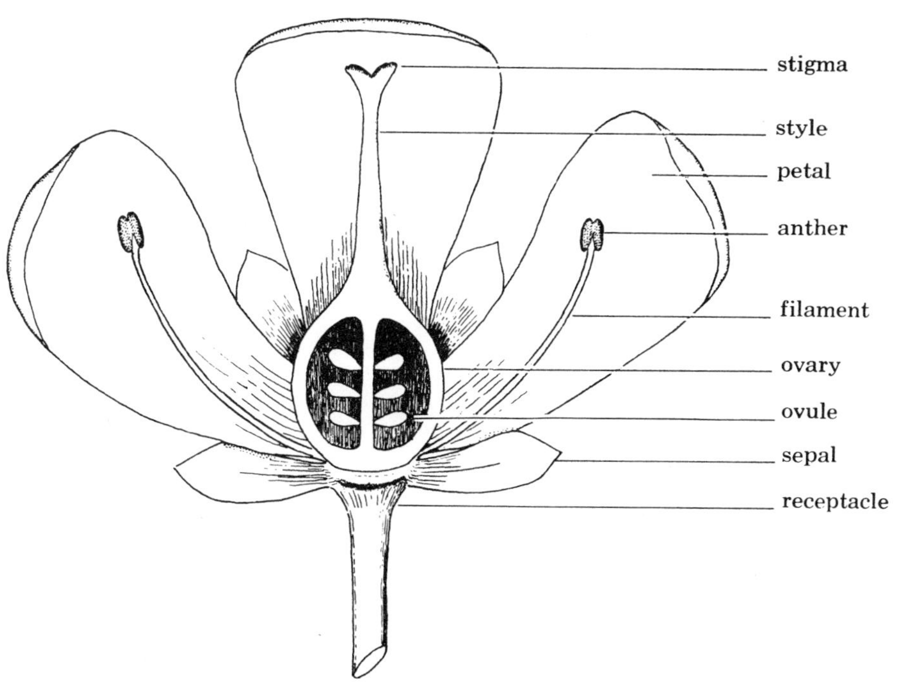

Parts of a flower

Glossary

Achene. A small, dry, hard, indehiscent, 1-seeded fruit.
Acicular. Needlelike.
Actinomorphic. Exhibiting radial symmetry as a regular flower.
Acute. Tapering to a pointed tip with the sides nearly straight.
Adaxial. On the side toward the axis.
Adnate. Grown together; applied to unlike parts.
Alpine. Land above the trees.
Androecium. The stamens and associated structures considered collectively.
Annual. From seed to maturity and death in one year or growth season.
Annulus. A single row of thick-walled cells which forms a ring around a fern's sporangium.
Anthesis. The period of blooming during which a flower is fully expanded.
Apical. At the top or tip.
Apiculate. Terminating abruptly in a small, generally sharp point.
Apomixis. The subsititution of some form of asexual reproduction for the sexual process, usually setting seed without fertilization.
Aquatic. Growing in water.
Areole. A small, clearly bounded area on a surface.
Aristate. Tipped with an awn or bristle.
Auricle. A small, projecting lobe or appendage, usually at the base of an organ.
Awn. A terminal, slender bristle on an organ.
Axil. Belonging to, or situated in, the axis.

Banner. The broad, erect, upper petal of the flower of legumes.

Barbellate. Minutely barbed.

Basifixed. Attached by the base.

Beak. A prolonged slender projection on the thicker organ such as a fruit, seed or petal.

Berry. A pulpy, indehiscent fruit with no true stone.

Bidentate. Two-toothed, important in *Carex* and some grasses.

Biennial. From seed to maturity and death in two years.

Bifid. Cut or divided into two lobes or parts.

Bilabiate. Two lipped (calyx or corolla).

Blade. The flat expanded portion of a leaf or petal.

Bract. A reduced leaf subtending a flower, usually associated with an inflorescence.

Bracteole. A small bract.

Bristle. A stiff, strong hair.

Budding. An asexual way of reproduction, usually the formation of an underdeveloped shoot or stem.

Bulbil, Bulblet. Minature bulbs.

Caespitose. Growing in clumps.

Callus. Hardened, downward extension of base of lemmas (actually part of the rachilla).

Campanulate. Bell-shaped.

Canescent. Covered with dense, fine grayish-white hairs.

Capitate. Head-like; collected into a dense cluster.

Capillary. With the form of a hair.

Capsule. A dry, dehiscent fruit composed of more than one carpel.

Carpel. A simple pistil, or one of the units forming a compound pistil.

Caryopsis. The grain or dry fruit of grasses.

Catkin. A scaly, deciduous spike.

Caudex. The woody base of an otherwise herbaceous perennial.

Caulescent. With an obvious leafy stem.

Cauline. Borne on the stem.

Chaff. Thin, dry scales.

Ciliate. Marginally fringed with hairs.

Circinate. Coiled from the top downward with the apex as a center.

Circumboreal. Distributed around the top of the world in the boreal (northern) region.

Circumpolar. Occurring around the pole, as of arctic plants mostly confined to far northern latitudes.

Circumscissile capsule. A capsule that dehisces by a cap coming off the top.
Clavate. Shaped like a club or baseball bat.
Claw. The narrow stalk of some petals.
Cleft. Cut about halfway to the midrib.
Cleistogamous flower. One which sets seeds even though it never opens.
Coalescent. Parts of one kind that have grown together.
Compound. Having two or more similar parts in one organ.
Conical. Having the shape of a true cone.
Connate. Grown together or attached, applied only to like organs.
Cordate. Heart-shaped with the notch at the base.
Coriaceous. Leathery in texture; tough.
Corm. A short, bulblike, underground stem.
Corymb. A flat-topped or convex, racemose, flower cluster, the lower or outer pedicels longer.
Cotyledon. A leaf of an embryo of a seed.
Couloir. A steep gorge or gully on the side of a mountain.
Crustacean. An aquatic arthropod of the class Crustaceae, typically having an exoskeleton.
Culm. Hollow or pithy, slender stems found in grasses and sedges.
Cuneate. Wedge-shaped, usually referring to the base of a structure.
Cyme. A flat topped or convex, paniculate, flower cluster with central flowers opening first.
Cymose. Arranged in cymes.

Deciduous. Falling off as leaves fall in autumn.
Decompound. Repeatedly compound.
Decumbent. Reclining or lying on the ground with stem tips ascending.
Decurrent. With an adnate wing or margin extending down the stem or axis below the point of insertion.
Decussate. Opposite pairs (usually leaves), alternating at right angles with those above or below.
Dehiscent. Opening spontaneously when ripe to discharge the contents.
Deltoid. Shaped more or less like an equilateral triangle.
Dentate. Having the margin cut with sharp teeth not directed forward.
Diadelphous. Stamens united by their filaments into two sets.
Dichotomous. Repeated branching in pairs.

Didymous. Developing in pairs.

Dimorphic. Having two forms.

Dioecious. Having staminate and pistillate flowers on different plants.

Disarticulate. To separate or break apart.

Disk. An outgrowth from the receptacle, surrounding the base of the ovary; in the Asteraceae, the central part of the head, composed of tubular flowers.

Dissected. Deeply divided into numerous fine segments.

Dolabriform. Refers to a hair shaped like a pick, attached toward the middle.

Dorsal. Pertaining to or located on the back.

Drupe. A fleshy one-seeded indehiscent fruit containing a stone; a stone fruit such as a plum.

E-, Ex-. Latin prefix meaning without, out of, from.

Eligulate. Without a ligule.

Elliptic. Longer than wide, widest in the middle and tapering toward the tip.

Emergent. Rising out of the water above the surface.

Entire. Without teeth on the margins.

Epigynous. A flower in which the hypanthium or the perianth is attached to the upper part of the ovary, the ovary then appearing inferior in position.

Epipetalous. Attached to the petals or corolla.

Equitant. Astride as if riding, such as the leaves of an Iris.

Erose. With an irregular margin, as if chewed by uneven teeth.

Evergreen. Persistent two or more growing seasons.

Exserted. Projecting beyond the enclosing organ.

Fascicle. A close cluster or bundle of flowers, leaves, roots, etc.

Fertile pinnule or **frond.** Leaf or portion of a leaf bearing spores.

Fibrillose. Composed of or breaking down into fibrils or fibers.

Filiform. Threadlike.

Fimbriate. Fringed (with longer or coarser hairs as compared to ciliate).

Flaccid. Lax and weak.

Floccose. Covered with loosely spreading and more or less tangled, long hairs.

Floret. The individual flower of the Asteraceae and Poaceae.

Floriferous. Bearing flowers.

Foliaceous. Leaflike; used especially with sepals or bracts.

Follicle. A dry fruit of one carpel that splits on one side.

Fornices. Set of small scales or appendages in the tube or throat of the corolla.

Free central. A type of placentation in which the ovules are attached to a central stalk within the center of the ovary.

Frond. Leaf of a fern.

Funnelform. Shaped like a funnel, with gradually widened tube.

Galea. A hood or helmet-shaped petal, usually the upper lip of a zygomorphic corolla.

Geniculate. Bent abruptly, as a knee.

Gibbous. Abruptly swollen on one side, usually near the base.

Glabrous. Without hairs, smooth.

Glandular. With secreting organs which produce small droplets of secretion.

Glaucous. Surface covered with whitish substance, which can be rubbed off.

Globose. More or less spherical.

Glochidiate. Barbed at the tip.

Glumes. The pair of bracts at the base of a grass spikelet, subtending a floret or florets.

Glutinous. Covered with sticky secretion.

Gynaecandrous spike. Referring to *Carex;* a spike with both staminate and pistillate flowers, the staminate below the pistillate.

Habitat. The kind of locality in which a plant grows.

Hastate. Shaped like an arrowhead except that the basal lobes are more divergent.

Head. A dense cluster of sessile flowers arising essentially from the same point on the peduncle.

Helicoid. Refers to an inflorescence with the axis curved as a helix.

Herb. A plant without persistent woody stem at least above the ground.

Heterosporous. Producing two different kinds of spores, usually of unequal size.

Hirsute. Covered with long, rather stiff hairs.

Hispid. Pubescent with coarse, firm and often pungent hairs.

Homosporous. Producing only one kind of spore.

Hydrophyte. A plant adapted to an aquatic habitat.

Hypanthium. The fused bases of calyx, corolla and stamens.

Hypogynous. Borne on the receptacle below or free from the pistil; pertains to perianth and stamens.

Imbricate. Overlapping as shingles on a roof.

Imperfect flower. A flower which has stamens but not pistils, or pistils but not stamens, regardless of what other parts may be present or absent.

Incised. Cut rather deeply and sharply.

Indehiscent. Not splitting open.

Indurated. Hardened.

Indusium. In ferns, the epidermal outgrowth that covers the sorus.

Inferior. Lower or beneath. Inferior ovary, one that is fused to the hypanthium.

Inflorescence. The flower-cluster of a plant.

Innovation. Basal shoot of a perennial grass or sedge.

Insectiverous. Referring to a plant which traps insects, usually with modified leaves.

Internode. The portion of stem between two nodes.

Involucre. A whorl of bracts subtending a flower cluster as in the heads of Asteraceae.

Involute. Rolled, not flat.

Irregular. Showing a lack of uniformity; asymmetric, as a zygomorphic flower.

Keel. A prominent dorsal ridge, analogous to the keel of a boat; the two lower and united petals of a papilionaceous corolla.

Labellum. Lip; the odd petal of an orchid flower.

Labiate. Lipped.

Lacinate. Cut into narrow lobes or segments.

Lanceolate. Lance-shaped, broadest near the base and tapering toward the tip.

Lateral. On the side of a structure.

Leaflet. The ultimate unit of a compound leaf.

Legume. A one-loculed fruit of a simple pistil usually dehiscent along two sutures.

Lemma. The lower of the two bracts immediately enclosing the floret of a grass.

Lenticular. Shaped like a double-convex lens.

Ligule. The strap-shaped part of a ray flower in the Asteraceae; the thin collar-like appendage on the inside of the blade at the junction with the sheath in grasses.

Limb. Expanded portion of corolla or calyx above the tube or throat.

Linear. Long and narrow with parallel margins.

Lobe. A projecting segment of an organ, too large to be called a tooth or an auricle.

Locule. The chamber or compartment of an ovary.

Loculicidal. Referring to fruit dehiscence directly into the locule.

Lodicules. The 2 or 3 minute scales at the base of the stamens in grasses, representing the perianth.

Loment. A legume fruit conspicuously constricted between the seeds.

Mericarp. An individual carpel of a schizocarp.

-merous. Referring to the parts in a whorl of floral parts (a flower having 5 sepals, 5 petals, 5 stamens, and 5 carpels is 5-merous).

Mesophyte. Plant adapted to growth under ordinary moisture conditions; intermediate between hydrophyte and xerophyte.

Microphyll. Usually a small leaf without a leaf gap, characteristic of fern allies.

Midrib. The central axis or vein of the leaf blade.

Monadelphous. Stamens united by their filaments into a tube surrounding the pistil.

Monoecious. Having unisexual flowers, but the staminate and pistillate flowers borne on the same plants.

Mono-. Greek prefix meaning one.

Monomorphic. Having one form.

Montane. Pertaining to the mountains.

Moraine. A deposit of gravel, sand and clay left on the ground by a glacier.

Mucilaginous. Gummy or gelatinous.

Mucronate. Having a small and short abrupt tip.

Nectary. Any structure which produces nectar.

Node. The point of insertion of a leaf or leaves.

Nutlet. Any small and dry nut-like fruit or seed.

Ob–. Latin prefix meaning reverse or inversion.

Oblong. Longer than wide with nearly parallel margins.

Obtuse. Blunt, with the sides coming together at an angle greater than 90 degrees.

Orbicular. Essentially circular in outline.

Oval. Broadly elliptic.

Ovate. Egg-shaped, widest near the base.

Ovoid. Shaped like a hen's egg.

Ovule. An immature seed before fertilization.

Ovulate cone. The female cone of a gymnosperm that bears ovules.

Palea. Uppermost bract which encloses stamens and pistil in a grass flower, usually lacking a midnerve.

Palmate. With 3 or more lobes, leaflets or nerves arising from a common point.

Palmatifid. Cut palmately.

Panicle. A branched indeterminate inflorescence.

Papilionaceous flower. A flower having bilateral symmetry, consisting of a banner petal, two wing petals, and two partly united keel petals.

Pappus. The modified calyx of Asteraceae, usually composed of bristles or awns.

Parasite. A plant which derives its food or water chiefly from another living plant.

Pedicel. Stalk of a single flower or a grass spikelet.

Peduncle. Common stalk of a flower cluster.

Peltate. Shape like a shield.

Pendulous Hanging or drooping.

Perennial. A plant that lives more than two years.

Perfect flower. A flower having both stamens and pistil even though it may lack a calyx or corolla.

Perfoliate. Having the base completely surrounding the stem.

Perigynium. A special bract which encloses the achene or ovary of *Carex*.

Perigynous. Referring to the union of perianth and stamens into a basal cup (hypanthium) distinct from the ovary.

Petiole. A leaf stalk.

Pinnae. Primary lateral divisions of a pinately compound leaf.

Pinnatifid. More or less deeply cut in a pinnate manner.

Pinnule. An ultimate leaflet segment of a fern leaf which is 2 or more times compound.

Placentation. Mode of attachment of the ovule to the ovary.

Planoconvex. Shaped like a lens that is flat on one side and convex on the other side.

Plumose. Feathery, often in reference to the style.

Polygamous. Plants with perfect and imperfect flowers.

Polyploid. With three or more sets of chromosomes (genomes) within the cells.
Pome. Fruit type with a core, like an apple.
Prostrate. Flat on the ground.
Pruinose. Strongly glaucous like a prune.
Puberulent. Minutely pubescent; covered with minute hairs.
Pubescence. Hairs of any type on the surfaces of plants.
Punctate. Dotted, generally with small pits that may be glandular

Raceme. An elongated inflorescence with a single main axis and single stalked flowers.
Rachilla. Axis of a grass spikelet; secondary axis; floret bearing axis.
Rachis. A main axis, such as that of a compound leaf.
Ray. The ligule or ligulate flower in the Asteraceae.
Receptacle. The end of the stem to which the other parts of a flower are attached.
Reflexed. Bent backward.
Regular flower. A flower which has radial symmetry; all parts are similar in size, shape and orientation.
Reniform. Kidney-shaped.
Reticulate. Net-like.
Retrorse. Directed backward or downward.
Revolute. Having the margins rolled back or under.
Rhombic. A shape with the widest axis at midpoint of structure, and with straight margins.
Rhizome. A creeping underground stem.
Rotate. Wheel-shaped, with short tube and wide limb at right angles to the tube.
Rugose. Wrinkled.

Saccate. In the shape of a sac or pouch.
Sagittal plane. Longitudinal plane drawn through a bilaterally symmetrical flower so as to divide it into right and left halves.
Sagittate. Arrowhead-shape.
Salverform. Trumpet-shaped; with a slender tube and an abruptly spreading limb.
Saprophyte. A plant which lives on dead organic matter. The term is often used to include the taking of food via fungi.
Scabridulous. Minutely roughened.
Scabrous. Rough to the touch.
Scape. A leafless flower stalk arising from a cluster of basal leaves.

Scapose. Bearing a scape.

Scarious. Dry, thin, scale-like, not green.

Schizocarp. A fruit which splits into its separate carpels at maturity.

Secund. With flowers or branches all on one side of the axis.

Septate. Having a partition.

Septicidal. Referring to dehiscence of fruit directly into ovary septum or partition.

Sericeous. Silky appearance due to long, soft appressed hairs.

Serrate. Having sharp teeth pointing forward.

Sessile. Attached directly by the base, without a stalk.

Setose. Covered with bristles.

Sheath. The basal portion of a leaf of a grass-like plant which surrounds the stem.

Silicle. A fruit similar to a silique, but not much longer than wide.

Silique. An elongated capsule in which the two valves or halves separate from the seed bearing partition.

Sorus. A cluster of sporangia.

Spatulate. Rounded above and narrowed to the base.

Spicate. Arranged in a spike.

Spike. A racemose type of inflorescence with sessile flowers.

Spikelet. Small spike; in sedges and grasses, one of the ultimate flower clusters consisting of glumes and the enclosed florets.

Spinose. An acuminate tip which is tough and stiff.

Sporangiophore. A branch bearing sporangia.

Sporangium. A sac-like structure containing spores.

Sporocarp. A specialized, swollen leaf bearing sporangia on the inside.

Sporophyll. A modified leaf which bears or subtends one or more sporangia.

Spur. A hollow appendage of the corolla or calyx.

Staminate flower. A flower with one or more stamens, but no pistil.

Staminode. A modified, non-functioning stamen which does not produce pollen.

Stellate. Star-shaped, referring to hairs with several branches from the base.

Sterile. Unproductive or infertile.

Stipe. Stalk of some pistils or fruits above the base of the perianth; any stalk.

Stipulate. Provided with stipules.

Stipule. A papery to leaf-like appendage at the base of the petiole of a leaf.
Stolon. An elongate, creeping stem on the surface of the ground.
Strobilus. A cone-like structure containing sporangia.
Stylopodium. An enlargement or disk-like expansion at the base of the style.
Subtend. To be directly below and close to.
Succulent. Fleshy and juicy.
Sub-. A Latin prefix meaning under, or almost.

Tendril. A slender, coiling or twining organ associated with stems.
Tepal. A sepal or petal, or a member of an undifferentiated perianth.
Ternate. In threes.
Terete. Round in cross-section.
Tomentose. Covered with tangled, matted or woolly hairs.
Truncate. With the apex or base transversely straight as if cut off.
Tubercle. A small swelling or projection.
Tuberous. Fleshy roots resembling stem tubers.
Tundra. Describes an area, a kind of vegetation, beyond treelimit.

Umbel. Usually a flat-topped inflorescence in which all the pedicels arise from the same point on the stem.
Umbelliform. With the form but not the structure of an umbel.
Uncinate. Hooked at the tip.
Urceolate. Urn-shaped, contracted at or just below the mouth.
Utricle. A bladdery, indehiscent, 1-seeded fruit.

Valvate. Having margins of adjacent structures touching at edges only.
Venation. The pattern of veins in a leaf.
Ventral. Pertaining to or located on the front or belly side.
Vernation. Arrangement of leaves in the bud.
Versatile. Referring to the mid-point attachment of the anther to the filament.
Verticillate. Arranged in whorls.
Villous. Covered with long, soft, often curved hairs.
Viviparous. Sprouting or germinating on the parent plant, as the bulblets in the inflorescence of some plants.

Weed. A plant which aggressively invades disturbed sites or habitats where man does not want it.

Whorl. A series of 3 or more similar structures radiating from a common point.

Wing. A thin flat projection from the side or tip of a structure; one of two lateral petals in a papilionaceous flower.

Xeromorphic. Having the form or appearance of a xerophyte.

Xerophyte. A plant adapted to life in dry places.

Zygomorphic flower. An irregular flower, having bilateral symmetry.

Selected References

Beetle, A. A., and M. May. 1971. Grasses of Wyoming. Univ. Wyo. Agr. Exp. Sta. Res. Journ. 39.

Davis, R. J. 1952. Flora of Idaho. W. C. Brown Co., Dubuque, Iowa.

Dorn, R. D., and J. L. Dorn. 1972. The ferns and others Pteridophytes of Montana, Wyoming, and the Black Hills of South Dakota. Available from the authors. University of Wyoming, Laramie, Wyoming.

Dorn, R. D. In manuscript. Manual of the vascular plants of Wyoming.

Hermann, F. J. 1970. Manual of the Carices of the Rocky Mountains and Colorado Basin. Agricultural Handbook 374, Forest Service, USDA.

Hitchcock, A. S. 1951. Manual of grasses of the United States. 2nd edition revised by Agnes Chase. USDA miscellaneous Pub. No. 200.

Hitchcock. C. Leo, and Arthur Cronquist. 1973. Flora of the Pacific Northwest. University of Washington Press, Seattle.

Oswald, E. T. 1966. A synecological study of the forested moraines of the valley floor of Grand Teton National Park, Wyoming. Ph.D. thesis, Montana State University, Bozeman, Montana. 101 pp. Unpublished.

Porter, C. L. 1962-72. A flora of Wyoming. Parts I-VIII. Univ. Wyo. Agr. Exp. Sta. Bulletins 402, 404, 418, 434; Res. Journals 14, 20, 64, 65.

Reed, J. F. 1952. Vegetation of Jackson Hole Wildlife Park, Wyoming. Amer. Midl. Nat. 48:700-729.

Scott, R. W. 1966. The alpine flora of Northwestern Wyoming. M.S. thesis, University of Wyoming, Laramie, Wyoming. 219 pp. Unpublished.

Shaw, R. J. 1958. Vascular plants of Grand Teton National Park. Amer. Midl. Nat. 59:146-166.

———. 1968. Vascular plants of Grand Teton National Park, Wyoming. Revised. Sida 4:1-56.

Weber, William A. 1972. Rocky Mountain Flora. Colorado Associated University Press, Boulder, Colorado.

Index

This is an index of family, generic and common names. Since specific epithets are arranged alphabetically, they have not been included (except for specific illustrations). Illustrations are referred to in boldface type..